1.0

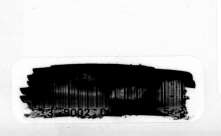

Dictionary of Water
and
Water Engineering

To June

Dictionary of Water
and
Water Engineering

A. NELSON, Dip. Min., M.I.Min.E.
(now Fellow), F.G.S.,
and
K. D. NELSON, E.D., B.Sc., M.I.C.E., M.I.E.
Aust.

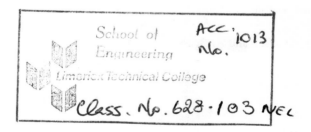
LONDON   BUTTERWORTHS

THE BUTTERWORTH GROUP

ENGLAND
Butterworth & Co (Publishers) Ltd
London: 88 Kingsway, WC2B 6AB

AUSTRALIA
Butterworths Pty Ltd
Sydney: 586 Pacific Highway, NSW 2067
Melbourne: 343 Little Collins Street, 3000
Brisbane: 240 Queen Street, 4000

CANADA
Butterworth & Co (Canada) Ltd
Toronto: 14 Curity Avenue, 374

NEW ZEALAND
Butterworths of New Zealand Ltd
Wellington: 26-28 Waring Taylor Street, 1

SOUTH AFRICA
Butterworth & Co (South Africa) (Pty) Ltd
Durban: 152-154 Gale Street

First published in 1973

© A. Nelson and K. D. Nelson, 1973

ISBN 0 408 00090 2

Printed in Hungary

Preface

From the earliest days of Man's history, water has influenced his destiny. Today, with the world's population expanding at a rate never before experienced, the need to develop and conserve all forms of water resources has become a matter of extreme urgency.

With world demand for water expected to double within the next twenty years, the lack of water resources is rapidly becoming one of the major problems facing both developed and undeveloped countries.

Coinciding with this upsurge in demand for water, there has been a terminology explosion which has created many difficulties in communications both within and without the engineering profession. With this problem in mind, we have attempted to compile a dictionary which will meet the needs of students and practising engineers. It should also prove of value to contractors, technicians, and laymen associated with water engineering projects.

Terms cover the many facets of water engineering and include urban and rural water supplies, irrigation, river improvement, harbours and sea defences, drainage, erosion, groundwater exploration, hydrography, flood protection, hydraulic machines, dams, and water power. References are also made to the related sciences of hydraulics, hydrology, hydrogeology, and meteorology.

In many instances we have felt obliged to explain rather than purely define terms, and in order to facilitate rapid linking up of associated and alternative terms cross-references have been included.

We have been aware of the inherent difficulty of co-ordinating some terms in the many English-speaking countries and have recorded what we consider to be the current usage; in certain doubtful cases we have expanded the definition.

A dual system of units, SI and imperial, has been introduced. In some instances an approximate transformation has been considered appropriate; however, where accuracy is essential, the conversion has been more precise.

During the preparation of the dictionary, we have consulted numerous books, technical journals, transactions, and other literature from British, Australian, and American sources. The very large number of references has made individual acknowledgements impracticable.

A. NELSON
K. D. NELSON

Note on Unit Abbreviations

The following abbreviations for compound units are used in the dictionary:

cumecs   cubic metres per second
cusecs   cubic feet per second
m.g.   million gallons
m.g.d.   million gallons per day
m.g.h.d.   million gallons per head per day

A

abandoned shore line A sea shore may be abandoned by depression of the sea or by uplift of the land. A lake shore may be abandoned by lowering of the entire lake by evaporation or drainage, or by differential tilting of the land, resulting in the exposure of the lake floor at one end or the other. See COAST OF EMERGENCE.

ablation The waste of a glacier by melting; the removal of surface rocks by water action.

Abney level A hand instrument for taking angles or levels along steep slopes; frequently used for the preliminary rough levelling for small irrigation and water supply schemes. See CLINOMETER.

abrasion The action of natural agents, particularly flowing water, in wearing away soils and rocks. The rock fragments may be roughly rounded and they abrade the beds of streams. See EROSION.

absolute density Weight per unit volume of particles (sand, soil, or gravel), excluding VOIDS. See DENSITY.

absolute drought Applied in U.K. to periods of 15 consecutive days or more to none of which is credited 0·25 mm (0·01 in) of rain or more. See BRITISH RAINFALL, PARTIAL DROUGHT.

absolute humidity See HUMIDITY.

absolute pressure The intensity of pressure above absolute zero. See GAUGE PRESSURE.

absorbed water Water held mechanically in a soil or rock and possessing physical properties which do not differ from ordinary water at the same pressure and temperature. See ADSORBED WATER.

absorbing well (or **well drain** or **drain well**) A well excavated or drilled mainly for drainage purposes. See DRAINAGE WELL, VERTICAL SAND DRAIN, WELLPOINT.

absorption (water) To suck in or imbibe water; the quantity of water which a rock or soil will incorporate, usually expressed in percentage terms of the original dry weight. A rock of low porosity can absorb or retain but little water, while a rock of high porosity may retain a large quantity. The figures show wide variations even for a rock of one type. Absorption is affected by soil texture, surface slope, rate of precipitation and air temperature. Water may be absorbed from rainfall or indirectly from streams, although normally groundwater moves towards the streams.

absorption, flood See FLOOD ABSORPTION.

absorption loss The quantity of water absorbed by the rocks and soils in the wetted area during the initial filling of a dam, pond, or canal. See SEEPAGE LOSS.

abstractors See DIVERTERS.

abutments (of dam) The supports for a dam which take a portion of the horizontal thrust; in the case of a concrete dam this thrust is taken by the rock sides of a valley. See ARCH DAM.

1

abutment wall See WING WALL.

Abyssinian tube well A pointed, perforated WELLPOINT which is driven into the water-bearing ground by ramming with a light pile hammer or by sledge hammer; water is extracted by pumping. Sometimes called NORTON TUBE WELL.

accelerated erosion The erosion of soil at a faster rate than that prevailing under normal or natural conditions; may be caused by excessive rainfall, or by human activities such as removal of root-bound surface layers of soil. See SHEET EROSION.

access manhole A MANHOLE provided on a large pipeline for cleaning and internal inspection. A convenient position is at a hollow or summit and near main valves.

access road A track constructed to enable plant, equipment, materials, and labour to reach the construction site. Frequently dams are built in isolated regions and consequently the provision of an access road can be important in cost and time.

accidental lake A lake formed in a drainage area in which the outlet has been blocked by debris, such as a landslide. The lake may be large or small and is frequently of short duration. See LANDSLIDE LAKE.

accident prone Certain persons have a predisposition to sustain more accidents than others when exposed to the same hazard. The ability of the water supply engineer to rapidly identify them is essential to minimising damage to the costly earthmoving equipment commonly used at construction sites such as dams.

accretion The building up of sediments from any cause as by stream action; the accumulation resulting from this process. See AGGRADATION.

acidification (or **acidising**) The use of acid, usually hydrochloric, to increase the water supply from a borehole which is failing owing to encrustations on screens and slotted pipes. It may also be used to improve the yield from bores in chalk or limestone by enlarging the fissures. See CALGON METHOD.

acid soil A 'sour' soil with a pH value below 7·0. See pH VALUE.

acid water Water containing traces of free sulphuric acid, mainly by contact with iron pyrites. Some subsurface waters are corrosive and contain a high proportion of solids, particularly the sulphates of iron. See TOXIC WATER.

acre A U.S. and British unit of area, frequently used to express the size of irrigation and catchment areas. Equal to 10 square chains or 4840 square yards (or 0·405 hectares).

acre foot The volume of water which will cover 1 acre 1 foot deep; equal to 43 560 ft^3 (1233·5 m^3). Unit of measurement for reservoir and pond capacity, also irrigation storages (domestic or town water supply reservoirs are usually measured in millions of gallons).

acre inch The quantity of water which will cover 1 acre 1 inch deep; equal to 3630 ft^3 or 22 600 gal or 102·8 m^3.

2

active layer The surface soil or deposit which undergoes volume changes with temperature, i.e. swelling when freezing and shrinking when thawing and drying. Also applied to the uppermost part of the PERMAFROST which thaws during milder weather and becomes waterlogged and forms MUD-FLOWS on sloping ground. Usually structural foundations are taken down below the active layer.

active method A construction method used in PERMAFROST areas. The frozen deposits are thawed prior to construction and kept thawed or removed; the foundation materials used are not subject to frost heave and settlement. See PASSIVE METHOD.

actual velocity See EFFECTIVE VELOCITY (GROUNDWATER).

adhesive water (or **pellicular water**) Water retained in a soil mass by molecular attraction; it forms a coating around the particles and may move from one particle to another. See INTERSTITIAL WATER.

adit Usually a rectangular heading or tunnel, horizontal or inclined, for tapping groundwater. It is often driven from a shaft and may be lined or unlined. Adits often form part of a well system in which wells and boreholes are sunk from the surface through the adits or from the floor of the adits. In some cases, the adits provide temporary storage for the water.

adsorbed water Water retained in a soil mass by physicochemical forces. It has physical properties which differ from chemically combined water, or mechanically held water at the same pressure and temperature. See ABSORBED WATER.

aeration, normal See NORMAL AERATION.

aeration zone See ZONE OF AERATION.

aerial photograph A photograph taken from aircraft at a definite height and making a series of parallel flights; used for map construction and geological interpretation; the photograph is usually vertical, i.e. taken by a camera pointing directly downward, although oblique photographs are used for some purposes. Frequently used to determine catchment size. See INFRA-RED PHOTOGRAPHY, MOSAIC, PHOTOGEOLOGY, PHOTOGRAM-METRIST.

affluent A tributary stream or system; a stream flowing into a river or lake.

afflux The variation between high-flood levels downstream and upstream at a weir; the rise of water level above normal level on the upstream side of a contraction or obstruction in a channel.

afforestation The planting of trees, shrubs, and cuttings of same on gathering grounds to retard runoff during and after heavy rains, or the melting of snow. Plantations reduce erosion and the transport of trash and debris into rivers and reservoirs. See TREE PLANTING.

A-frame groynes A modified form of the CRUCIFORM GROYNE in which two spars, with a third as crossbar, form 'A'-shaped frames instead of crosses. Green willow poles are planted against the groynes, on the upstream side, with their feet rooted in the river bed. These are small groynes

3

used for protection along low eroding banks and also as short training walls, under conditions with only minor floods.

age (of stream) The stage reached in a stream's development or growth. See GRADED STREAM, MATURE STREAM, OLD RIVER, PROFILE OF EQUILIBRIUM, YOUTHFUL STREAM.

aggradation See ALLUVIATION.

aggraded floodplain An ALLUVIAL FLAT which is often covered by floodwater of a MATURE RIVER. When the waters subside, the plain is covered with a new layer of sediment. See GRADED RIVER FLOODPLAIN.

aggregate The crushed stone, sand, or gravel which is bound together to make concrete. See CHIPPINGS.

agricultural drain (or **field tile**, **land drain** or **field drain**) Earthenware or porous concrete pipes of about 75 mm (3 in) internal diam., laid with open joints, end to end, for subsoil drainage. See FRENCH DRAIN, SUBSOIL DRAINAGE.

air, damp See DAMP AIR.

air, dry See DAMP AIR.

air-entrained concrete A concrete containing an agent to entrain about 3·5% of air. Sometimes used in mass concrete for dams and earth dam cutoffs, as it gives marked improvements in the workability and cohesion

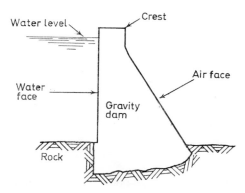

Figure A.1.   Air face (dam)

of the mix. In the U.S.A. and some other countries, air-entrained concrete is favoured on account of its superior resistance to freezing and thawing.

air face (dam) The downstream face or side of a dam (Figure A.1). See WATER FACE.

air-lift pump An appliance (Figure A.2) for lifting water in a well or sump. It consists of an air compressor at the surface and two pipes hanging vertically, with one pipe perhaps within the other. The smaller pipe delivers compressed air to the depth at which water occurs. At this point a nozzle discharges the air into the free water and by aerating it causes its density to drop. This mixture of water and air is then forced upwards by the head of groundwater. Large volumes of dirty or gritty water can be removed in this way but the consumption of compressed air is high.

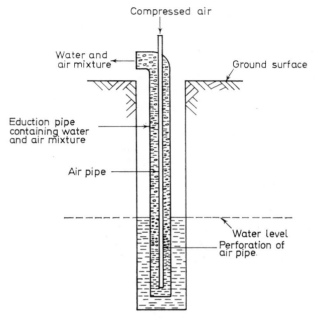

Figure A.2. Air lift pump

Air-lift pumps may be used to lift water into a reservoir or storage. See PULSOMETER PUMP, SUBMERSIBLE PUMP.

airline correction A correction required in measuring water depths to determine true depths. If large vertical angles are induced in log-lines by great depths, high velocities, inadequate sounding weight, or any combination of these, two separate corrections are required: (1) airline correction for the section of line above water surface and (2) wet line correction for the section under water.

air release valve A valve to release air or gas which tends to accumulate at high points on pipelines.

airspace ratio The ratio of (1) volume of water that can be drained, under the force of gravity, from a saturated soil sample to (2) total volume of voids.

5

air valve   A valve placed at the summit of a pipeline to allow air to escape but not water; any valve to discharge air when pipe mains are being filled or to admit air when they are being emptied.

air vessel   A small air chamber fixed to the pipeline on the discharge side of a RECIPROCATING PUMP to act as a cushion to minimise the pulsations of the pump.

algal control   The treatment of water with chemicals to suppress algae. The most widely used algicide is copper sulphate (bluestone) and the dose ranges between 0·05 to 10 p.p.m. (mg/l) depending on the type of organism.

alignment chart   See NOMOGRAM.

alkaline soil   A non-acid soil with a pH value above 7·0. See pH VALUE.

Allen's method   See SALT VELOCITY METHOD.

alluvial cone   A deposit of sediment similar to an ALLUVIAL FAN but with steeper slopes; material usually finer than that in a DEBRIS CONE.

alluvial deposit   A deposit of sand or gravel formed along lengths of a river where the water velocity has diminished. See COMPETENCE.

alluvial fan   A low fan-shaped heap of sediment deposited by a river when its gradient is suddenly diminished on entering a plain or open valley floor; deposits range from small heaps to large areas covering many square miles or square kilometres. Alluvial fans sometimes present engineering problems; culverts and bridges may be made useless, and roads or pipelines may be cut or deeply covered by debris.

alluvial flat   A deposit of mud and silt in the lower course of a mature river; the sediment may be up to 9·1 m (30 ft) or more thick.

alluvial river   A river which flows in a channel carved in its own deposits of alluvium. See REJUVENATION.

alluvial tract   See PLAIN TRACT.

alluviation (or **aggradation**)   The deposition or building up of sediments by stream or water action, particularly along rivers and estuaries where flow is retarded. See DEGRADATION.

alluvium   A fine-grained deposit, composed mainly of mud and silt, deposited by a river. See FLOODPLAIN, TERRACES.

altar   A step in the wall of a dry dock. When the dock is empty, the altar holds the feet of the wooden shores which keep the vessel steady.

altitude   The vertical height of a water surface, hydraulic point, or ground measured from a reference plane, Ordnance datum, or mean sea level. See BENCH MARK, ORDNANCE DATUM, STAGE.

alveus   The channel or bed of a river occupied by normal flows of water.

Ambursen dam   See FLAT-SLAB DECK DAM.

American caisson   See BOX CAISSON.

amplitude   (1) The depth of a wave measured from the level of calm water; extent of any oscillation or vibration. (2) The width of the meander of a river (Figure A.3).

anabranches   Small interlacing streams (Figure A.4). See BRAIDED COURSE.

analogy   A scientific comparison between two phenomena; an electrical

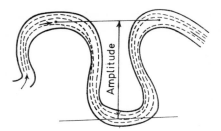

Figure A.3.   Amplitude

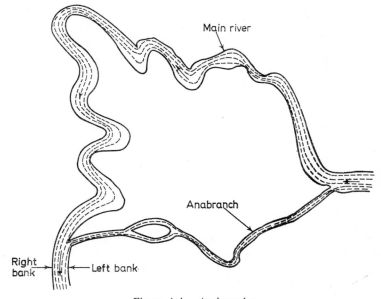

Figure A.4.   Anabranches

analogy has been applied to solve seepage problems, and alternating current is used as an analogy in the study of tides.

ancestral rivers   An Australian term for an ancient river system, usually of major dimensions. The limits of an ancestral river may be delineated by aerial photographs. The meander pattern of existing streams may change suddenly, indicating that they are flowing over an ancestral river course. See PRIOR RIVERS.

anchorages (pipes)   Thrust or anchor-blocks, usually of concrete, placed along a pipeline to prevent or reduce longitudinal stress being transmitted from pipe to pipe. They are often necessary on bends or when a main is on a gradient steeper than about 1 in 3. In firm ground any concrete required along the pipeline is keyed into the bottom and sides of the trench (Figure A.5).

anchorage spud   See SPUD.

anchor and collar   A strong metal hinge for lock gates; it is built into the

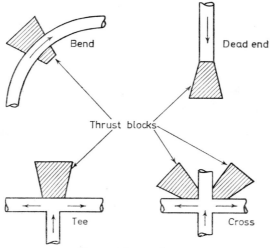

Figure A.5.   Anchorages

masonry or concrete of the lock and a hole in a projecting plate receives the pintle of the gate.

anchor buoy   A BUOY fixed in position by an anchor. See BELL BUOY.

anchored trees   (bank protection) The use of bushy growing trees, well anchored together, for protecting an eroding river bank; rapid and inexpensive where suitable trees are available locally. The trees are laid along the bank, butts upstream, and anchored by wire ropes or cables to posts or buried logs set well back from the bank. To prevent any loose trees from floating or moving at high water, they may be weighed down by concrete blocks or stonemesh sausages. Willows and vegetation are often planted behind the trees. The method has been used widely in south-eastern Australia and in New Zealand. See CRIBWORK, UNDERWATER APRON.

anchor-ice   A layer of ice up to about 50 mm (2 in) thick on the bed of a river, sometimes formed during clear cold nights; it usually becomes free and floats to the surface after sunrise.

anemometer   An instrument for measuring total wind travel in a given time or for continuously recording the velocity and direction of wind movements. See BEAUFORT SCALE.

8

angle of repose The maximum slope, measured from the horizontal, at which rocks, soils, and loose material on the banks of canals, rivers, dams, or embankments remain stable. See BATTERS.

angle of slide The angle (measured in degrees from the horizontal) of the sloping slides of a canal, reservoir, or other cutting, at which a LANDSLIDE will start. The angle varies considerably and the water content has a major influence. See STABILITY NUMBER.

angularity correction The correction made to the observed velocity when the direction of water flow is not exactly at right angles to the DISCHARGE SECTION LINE. See MODIFIED VELOCITY.

angular stone Broken stone obtained from the quarry, as distinct from gravel, boulders, or trimmed stone; may be large and used for groynes, etc. or small as used for concrete aggregate. See STONEMESH CONSTRUCTION, STONE RIPRAP.

anhydrous Without water, especially WATER OF CRYSTALLISATION.

annual depletion rate The average rate at which withdrawals deplete storage in reservoir of groundwater during a period of one year.

annual flood The maximum momentary discharge measured in a WATER YEAR.

annual flood peak series See PARTIAL DURATION SERIES.

antecedent moisture The moisture content of a soil at the commencement of a runoff period. It is often given as an index determined by summation of weighed daily rainfalls for a specified period (from 10 to 20 days) before the runoff considered.

antecedent stream A stream that deepens its valley and maintains its course in spite of slow tilting of the ground. See DEFEATED STREAM.

antecedent wetness See ANTECEDENT MOISTURE.

antiflood valve A CHECK VALVE so positioned in drainage channels that water discharges at low levels and so prevents flooding.

antiseep diaphragm See CUTOFF COLLAR.

antitranspirants Monomolecular films of high alcohols which are introduced onto the open water surface of storages to reduce the EVAPORATION losses.

apparent velocity (groundwater) The apparent rate of motion of groundwater in ZONE OF SATURATION; may be expressed by

$$V = Q/A$$

where Q is the volume of water passing through a cross section of area A in unit time. See EFFECTIVE VELOCITY.

approach channel See CHANNEL OF APPROACH.

approach velocity The mean velocity of water measured in the CHANNEL OF APPROACH.

appropriated rights (water) An individual's rights to the exclusive use of water, based strictly on priority of appropriation and the beneficial use of the water, and without limitation of the place of use to RIPARIAN LAND. See WATER RIGHTS.

apron A hard floor or surface to the bed or banks of a river or canal, or below chutes, spillways, groynes or the toes of dams, to minimise scour. It may consist of bagwork, mattress, timber, mass concrete, riprap, or preferably reinforced concrete. See STILLING POOL.

apron, falling See UNDERWATER APRON.

aqueduct A generic term for works including pipes, tunnels, open or covered channels, canals, and elevated structures for the bulk conveyance of water from source or storage to area of consumption. See TRUNK MAIN.

aquiclude A deposit or mass of low porosity which absorbs water slowly but does not transmit it freely enough to form useful supplies for a well; any saturated deposit, containing an interconnected system of interstices, but which does not yield useful supplies of water.

aquifer (Lat. *aqua*, water, and *fero*, I bear) A relatively permeable or fissured deposit which yields useful supplies of water when tapped by a well. In the U.K. the principal aquifers are the Chalk and Triassic sandstones, together with the Lower Greensand and the Oolites. See ARTESIAN AQUIFER, ARTESIAN BASIN.

aquifer, recharge of See GROUNDWATER DECREMENT, GROUNDWATER INCREMENT, INTAKE AREA OF AQUIFER.

aquifuge A relatively impervious rock which has no interconnected fissures, voids, or openings and hence cannot absorb or transmit water. See IMPERVIOUS ROCK.

Arany's number A number indicating the moisture content of a soil at which the material just starts to flow and to behave as a fluid. It is determined by pulling a thread of the soil out with a pestle as water is mixed into it. The moisture is at Arany's number when the thread starts to bend.

arch dam A dam which is curved in a horizontal plane, convex upstream. The water load is transmitted horizontally by arch action to the valley or canyon walls and it must therefore be built on solid rock, as yielding ground could cause instability and failure. The construction material is commonly concrete, although brick or stone is sometimes used. (See MULTIPLE-ARCH-TYPE DAM.)

Archimedes's principle A principle credited to Archimedes (287–212 B. C.), a celebrated Greek mathematician and engineer. It states that when a body is wholly or partly immersed in a fluid, the total uplift on the body is equal to the weight of the fluid displaced.

area, catchment See CATCHMENT AREA.

area curve A curve showing area of cross section, or other area of a stream, channel, or structure as a function of elevations.

area, discharge See DISCHARGE AREA.

area, drainage See CATCHMENT AREA.

area of aquifer (intake) See INTAKE AREA OF AQUIFER.

area of artesian flow   The area overlying an aquifer in which the water is under sufficient pressure to rise above the surface. See ARTESIAN PROVINCE.

area of groundwater discharge   The surface area embracing all points at which groundwater is discharged as springs or as evaporation from soils and plants.

area of influence (well)   The surface area around a well or borehole with the same horizontal extent as that part of the underlying water-table that is lowered by pumping from the well at a given rate of discharge. See CONE OF DEPRESSION.

areic area   An area with no flowing streams. See DRY VALLEY.

arêtes   See COMB RIDGES.

arid   Dry, parched; climate or land where agriculture requires irrigation because precipitation is inadequate. See SEMI-ARID.

arid cycle   See CYCLE OF EROSION.

arroyo   A narrow valley or gully, usually with steep banks, in an arid region; located above the water-table and often dry owing to infrequent rainfall or cloudbursts.

arterial drain   A MAIN DRAIN which has secondary drains leading into it.

artesian aquifer (or **confined aquifer**)   An aquifer in which the water is under pressure and confined beneath impermeable deposits.

artesian basin   An ARTESIAN AQUIFER in the form of a broad basin. The regional artesian basin of Queensland, Australia, covers some 1 550 000 km² (600 000 sq. miles) and supplies water to many deep wells in arid country. The Chalk of the London Basin is another example. Sands and sandstones form the principal source of artesian and other water. Limestone deposits may yield water under favourable conditions.

artesian discharge   The flow of water from an ARTESIAN WELL or ARTESIAN SPRING; the volume of water thus discharged.

artesian flow area   See AREA OF ARTESIAN FLOW.

artesian head   The PRESSURE HEAD of ARTESIAN WATER.

artesian head, negative   A term sometimes applied to a well in which the HYDROSTATIC PRESSURE is negative and the free water level is below the existing water-table level.

artesian head, positive   Applied to an aquifer or well in which the static pressure is positive and the free water level is above the existing water-table level. See POSITIVE CONFINING BED.

artesian pressure   The pressure of the water in an ARTESIAN WELL.

artesian province   A region within which artesian aquifers occur and where geological conditions are in general similar. Although a general similarity exists, the water yield and its quality may vary from point to point.

artesian slope   Refers to an ARTESIAN AQUIFER which slopes or dips beneath impermeable strata. Water seeping down from the outcrop into the aquifer at depth is stored there under pressure beneath the impervious cover. The structure is known as an ARTESIAN SLOPE.

artesian spring (or **blow-well**)   An uncommon type of SPRING in which the ARTESIAN WATER emerges along a fracture or opening in the impervious deposit overlying the aquifer.

artesian water   Groundwater which is under pressure and confined beneath relatively impermeable deposits. The water pressure would cause it to rise above the level at which it is encountered in drilling.

artesian well   A well which has tapped artesian water; a well in which the hydrostatic pressure of the water is sufficient to cause it to rise in the well and sometimes reach the land surface.

artesian well (or **free flowing bore**) (drainage)   A well put down for tapping shallow water, under artesian or near-artesian conditions, for drainage and not for usage. A typical installation would be a 100 mm (4 in) diam. high-density polythene pipe, the lower end being slotted to allow the entry of water which at the surface is discharged via a 'T' piece to a deep drain. In favourable localities, batteries of such wells are used to lower the water-table over considerable areas. See WELLPOINT.

artificial catchment   A small catchment which has been treated, wholly or in part, to increase the runoff. In the IRON-CLAD CATCHMENT one section of the area is covered with flat iron sheets. Some form of bitumenous preparation is another method of covering the ground. Another method is known as the ROADED CATCHMENT which involves the compaction of the area by rollers after it has been subdivided into a series of roads. The roads are about 7·3 m (24 ft) wide with a cross-fall from crest to drain of about 1 in 20. The table drains have a slope of approx. 1 in 200.

artificial harbour   A sheltered area of sea formed by building BREAKWATERS (see NATURAL HARBOUR).

artificial recharge   Augmenting the natural infiltration of surface water or rainfall into underground rocks or channels. The recharge may be by spreading of water, by changing the natural conditions, by some form of construction, or by putting down wells. See INVERTED WELL.

artificial storages (streams)   Measures to reduce the irregularity of stream flow and in which the excess water is temporarily stored by the erection of dams and similar works. The water so stored may be used for town supplies, power, or irrigation. A dam may be used wholly or partly for the storage of floodwater and so reduce the flooding of lowlands along a ˉiver. See NATURAL STORAGES.

asbestos-cement pipes   Pipes made of a mixture of asbestos and Portland cement which can withstand pressure and also internal and external corrosion. They were first manufactured in the U.K. in 1928 by Turners Asbestos Cement Co. They are now used by many water undertakers; in 1933 a British Standard was issued, revised in 1956 to cover pipes from 50 mm to 0·6 m (2 to 24 in) diam.

ashlar   Freestones as they come from the quarry; stones cut square and dressed. Ashlar facing is often used in masonry dams.

Ashley type groynes   See IMPERMEABLE GROYNES.

aspen A variety of poplar often used for soil conservation and river works; best propagated from root cuttings.

astyllen A low barrier or dam placed across an adit to retard the outflow of mine water.

Atlantic type coastline A coastline in which faulting, and not folding, plays a dominant role, and its line is not parallel to the trend of the mountain chains (see PACIFIC TYPE).

atmidometer (or **evaporimeter** or **atmometer**) An instrument for measuring EVAPORATION (q.v.).

atmosphere (1) The gaseous matter surrounding the earth. Composition varies very slightly according to altitude and region; it is mainly nitrogen and oxygen, with traces of carbon dioxide, argon, helium, etc. In addition, air contains water vapour and dust particles in very variable amounts. (2) A unit of pressure. That pressure which will support a column of mercury 760 mm (29·92 in) high at 0°C, sea level and latitude 45°. 1 normal atmosphere = 101·325 kN/m² (14·72 lb/in²). See HUMIDITY.

atmospheric water The water present in the atmosphere; it may be gaseous, liquid or solid. See RAINFALL, WEATHER, CLIMATE.

atoll A roughly circular coral reef enclosing a LAGOON (Figure A.6).

attached groundwater That portion of groundwater retained on the surfaces of particles, against the force of gravity, during drainage or pumping. See SPECIFIC RETENTION.

augers Power- or hand-operated boring tools varying from about 38 mm to 0·6 m (1½ to 24 in) diam. Augers are often used to investigate the foundation conditions of earth dams and the availability of suitable construction material in the borrow pit area. See EARTH AUGER.

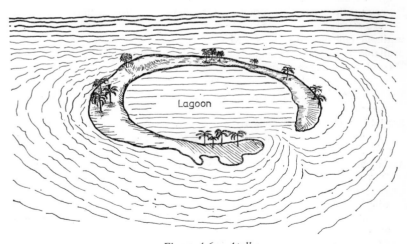

Figure A.6. Atoll

automatic irrigation An irrigation system which is automatically controlled by electrical, pneumatic, or mechanical devices.

automatic pump A pump provided with a FLOAT SWITCH which starts and stops the pump automatically depending on the water level in the suction sump or storage. Pumping stations are now often automatic in operation with indicators and recorders designed for remote reading.

automatic self-closing valves Valves fitted to large aqueducts which close automatically when the discharge exceeds a predetermined figure and so prevent possible damage to the pipeline. See FLOAT VALVES.

automatic siphon spillway See SIPHON SPILLWAY.

automatic water treatment plants Water treatment plants equipped with automatic cleansing filters and also possibly with automatic chemical equipment. All chemicals are delivered in bulk as liquids and stored in tanks which feed service tanks. The latter, in turn, are automatically kept full and the chemicals are also automatically mixed in the proper proportions inside the tanks.

automation A general term pertaining to the automatic control of machines, such as pumps and ancillary operations, by electronic control equipment, photoelectric cells, instrumentation, and remote control. Its aim is to reduce labour, attendants, and operational costs.

auxiliary protection (groynes) Works and practices to protect groynes from damage and premature failure and also to reduce maintenance costs. Various types of mattresses and aprons are used to reduce underscour and may be placed at the root and head of the groyne. At the root, aprons may be carried up the bank above flood level. At the head, protection is provided on the stream bed round the bend where flow is swift. These aprons and mattresses are made flexible to maintain contact with the ground. See APRON, MATTRESS, PILOT CHANNELS, SINKER GROYNES, STONE-MESH MATTRESS.

available moisture capacity (A.M.C.) See AVAILABLE SOIL WATER.

available soil water The soil water which is available or can be taken up by plants at any moisture content between FIELD CAPACITY and PERMANENT WILTING POINT. Sandy soils contain less available water than clay soils and loams form an intermediate group. Also called AVAILABLE MOISTURE CAPACITY.

available water-holding capacity (soils) The difference between the percentage of water at field capacity and the percentage of water at wilting point. See WILTING COEFFICIENT.

avalanche A large slide of ice, snow, and debris down a mountain slope. See LANDSLIDE.

average annual rainfall For a specified place, the average annual value for rainfall is the mean of the annual amounts over the period for which rainfall statistics are available. In the U.K. the period seldom exceeds 40 years.

average velocity (groundwater) The mean distance covered by mass of

14

groundwater per unit of time; equals total volume of groundwater passing through unit cross-sectional area per unit of time, divided by porosity of material passed through. See APPARENT VELOCITY.

avulsion (1) The sudden removal of land by flood; the accelerated erosion of shore by storm waves. (2) To sever or divide an area of land by a stream cutting across the narrow neck of a horseshoe bend. See OXBOW LAKE.

axial-flow pump A propeller pump in which the water flow is axial. The principle is similar to that of a ship's propeller and the pump is used for lifting large quantities of water over small heads. See PUMP.

axis of channel A line joining the mid-points in successive cross sections of a channel (see THALWEG).

axis of dam See DAM, AXIS OF.

B

Bacillus coli A rodlike germ sometimes found in drinking water and often derived from sewage pollution.

back acter An EXCAVATOR fitted with front-end equipment comprising a jib with an arm and bucket. Although designed primarily for vertically sided trenching, it is also useful for bulk digging below track level and for many excavations for drainage, ditching, and other hydraulic works.

backblowing A method of improving the water yield from boreholes, particularly in fissured rocks. It is not unlike SWABBING in action and is sometimes more effective. It employs compressed air (1) for pumping in the bore by air lift, or (2) by pumping in air (when the top of the hole is sealed) until the maximum pressure is obtained and then releasing it suddenly. See SHOT FIRING.

back cutting The extra excavation carried out to make up a canal or other hydraulic embankment when the material available from other sections of the work is insufficient (see BALANCED EARTHWORKS).

back filling (or backfill) Material excavated at a site or construction and re-used for filling, as in groynes and trenches for pipelines, etc.

back-inlet gully A cast-iron or glazed stoneware branch entry to a drain; water-sealed and covered by a grating, but open to the air. It receives rainwater or discharge from waste pipes under the grating but above the water seal.

backshore The area of shingle or sand beach between normal and abnormal high-water marks. See FORESHORE.

backwash The back flow of water which follows the breaking of a wave; it scours and carries the finer material towards deeper water where it is temporarily deposited (see UNDERTOW).

backwater Water held back in a channel or stream by a dam, regulator, or obstruction; water stored during high tide and discharged during low tide. See AFFLUX.

backwater curve A curve representing the BACKWATER surface upstream

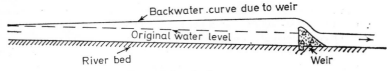

Figure B.1. Backwater curve

along an open channel from a dam or weir or obstruction; it may extend upstream for several miles (Figure B.1). See STREAM PROFILE.

bacteria bed (or **biological filter** or **contact bed** or **continuous filter** or **percolating filter**) A layer of FILTER MATERIAL such as clinker or rock which will expose sewage effluent to the oxidising action of micro-organisms in the air (see SLOW SAND FILTER).

badlands Denuded, rough, and irregular land with the surface containing many gullies and ridges (U.S.A.).

baffle-pier A wall, mass, or obstruction erected in the path of high-velocity water to dissipate energy and reduce scour; a pier on the apron of an overflow dam. See GROYNES.

baffles A series of plates, vanes, or guides set in a channel or conduit to secure more uniform flow conditions by damping eddies.

bagwork A form of bank protection used on rivers and sea walls. Small bags are filled with dry concrete and tamped by hand against the section to be protected. Gravel is sometimes used instead of concrete. Bagwork along sea walls is often held together by steel dowel rods driven through the bags. See REVETMENT.

bailer (or **sand pump** or **sludger**) A pump for removing sludge, or sand and water or mud, from a well or borehole. It consists of a tube or pipe with a valve at the lower end which opens upwards. An opening at the side at the upper end of the pipe enables the sediment water to be discharged.

Bailey bridge A military bridge of welded lattice construction developed in Britain during World War II by Sir Donald Bailey, since used extensively for water crossings in many countries. See MULBERRY HARBOUR.

bai-u Applied to the season with abundant rains in southern Japan and in parts of China (where it is termed MEIYU).

Baker bell dolphin See BELL DOLPHIN.

balance bridge See BASCULE BRIDGE.

balanced earthworks Earthworks in which the 'cuts' equal the 'fills'. The ideal is not easily achieved but is a target when planning the layout of trenches, water supply channels, and other hydraulic works.

balance storage See PONDAGE.

balancing tank See EQUALISING RESERVOIR.

band screen An endless moving band of wire mesh or perforated steel plates used at INTAKE WORKS to retain and remove solid matter from the water. The water passes through the screens and the solids are washed

16

off by water jets. The washings are collected in a launder and returned to the river or source well below the intake. See SCREENS, DRUM SCREEN.

bank (or **embankment**) The elevated ground along the sides of a stream or lake or channel; an elevation in the sea; a shoal; any ridge or earth slope formed or trimmed to a definite shape. See EMBANKMENT, RIGHT BANK.

bank, beaching of Forming a layer of stones along a river bank to prevent erosion and undercutting; may form part of other protective works, such as groynes and vegetation. See BEACHED BANK.

bankful discharge The DISCHARGE at BANKFUL STAGE.

bankful stage The stage when a stream fills its channel and just overflows its natural banks.

bank materials The soils and other superficial deposits used in the construction of embankments or earth dams. The material must be sufficiently impervious to reduce seepage to a safe rate and must also be sufficiently stable to permit economic side batters. Material containing about 25% clay and the remainder sand and silt makes good impervious banks. Too much clay weakens the bank and also makes it liable to volume changes with changes in moisture content, while insufficient clay can cause seepage. Bank materials are usually tested to determine their suitability before construction. See CLAY, SAND, TOP-SOIL COVER.

bank protection Measures to combat erosion along river and canal banks and along coastlines. They may include costly works of a permanent or semipermanent nature such as sheet piling, stone gabions, etc., or more temporary works which might last only a short period until vegetation has been established along the eroded banks. In some cases, bank protection to prevent erosion and to check MEANDER MIGRATION has helped to stabilise a river for three or four miles downstream. See GROYNES, MATTRESS, PERMANENT BANK PROTECTION, REVETMENTS, RIPRAP, SEA-WALLS, SLOPING OF BANKS.

bank, raw See RAW BANK.

bank revetment See REVETMENT.

bank sloping See SLOPING OF BANKS.

bank storage (or **lateral storage**) The water absorbed by the banks of a stream, reservoir, or lake at high water levels and returned to it, wholly or in part, as the water level falls. See CHANNEL STORAGE.

bar A barrier; a hard ridge of rock cutting across a river bed; a bank of gravel or silt at the mouth of a river or harbour and often unstable. Bars at the mouth of rivers may be formed in several ways: (1) Across the mouth of a tidal estuary it may be due to littoral or shore currents which drift material across the mouth of the estuary, or to eddies and still water produced by currents at the entrance. (2) Wind and waves may drive sediment across the mouth of a river entering a lagoon or bay of a tidal sea. (3) The checking of the current when sediment-bearing rivers enter inland seas or lakes often causes the load to be deposited as a bar.

17

barge A flat-bottomed boat used to carry heavy loads by canal or river. In the U.K. it has a beam of 4·3 m (14 ft) or more and the loads range from 70 to 150 t. On the continent of Europe loads up to 500 t may be transported.

barge bed A place where barges can moor and rest on a mud bottom at low tide near the bank of a river. A wall is usually built to retain the river bank. See CAMP SHEATHING.

Barnes's formula A U.K. formula used to determine velocity of flow v in ft/s in sewers:

$$v = 107m^{0·7} \sqrt{i}$$

where m is the HYDRAULIC MEAN DEPTH in feet and i is the slope of the sewer. See CRIMP AND BRUGES'S FORMULA.

barrage A low barrier or dam, incorporating a series of gates, built across a river to regulate its upstream level or for irrigation. A barrage may be erected to prevent ingress of salty sea water or to impound in the loop of a river tidal water used for scouring out the lower channel and thus maintaining its capacity.

barrier basin A basin formed by landslide debris in a narrow valley. A lake may be formed upstream from the obstruction. Other types of barrier basins include hollows behind morainic barriers, lagoons, and bays which have been closed in by BAY-MOUTH BARS, and depressions made by a sheet of lava which chokes a valley. See ICE GORGE.

barrier beach (or sand reef) A type of beach ridge formed a short distance offshore; sometimes occurs where the coast has a gentle slope and the waves break some distance out with deposition of material. Storm waves may build up the barrier above water level with the formation of a lagoon between the mainland and the barrier. In time the lagoon may become filled with sediment. See BEACHES.

bascule bridge (or balance bridge or counterpoise bridge) A river bridge hinged at the bank to enable ships to pass through by raising the portion over the navigable channel and lowering the portion over the bank behind the hinge. The construction material is usually aluminium alloy to minimise the weight to be moved by the machinery. Tower Bridge, London, is a well-known example in the U.K.

base course A layer of specified material and thickness placed on the subgrade (the prepared and compacted ground below a structure or road) to spread the load, minimise frost action, and provide drainage.

base flow The sustained flow of streams from snow and glaciers or from underground storage and not related to direct runoff; sustained flow resulting from drainage of large lakes or springs or other outflow of groundwater, as opposed to surface runoff. See PERENNIAL STREAM.

base-level The level of the body of water into which a river flows, usually the sea or a large lake. See TEMPORARY BASE-LEVEL.

18

base map (or **working map**) The map used during GEOLOGICAL MAPPING. A topographic contour map or aerial photographic map for plotting the geologic information obtained in the field. If used to investigate water resources, all porous and impervious beds are marked and also old or existing wells, springs, streams, etc.

base period When the BASE FLOW of a stream is exceeded as a result of heavy rainfall or a storm.

basic properties Soil properties such as consolidation (or compaction), shear strength, and permeability. These properties affect the behaviour of a soil at an engineering site. The results of consolidation tests are useful when the deposit is used as 'fill' material in earth dams or as foundation material. The shear strength and permeability of a soil are important in connection with dam foundations and earth dams and also when excavations are made in sands. See DAM, PERMEABILITY.

basin Topographically, either an area drained by a river and its tributaries or low-lying land encircled by hills. Geologically, an area in which the stratified rocks dip towards a central spot, and the rocks are said to possess a *centroclinal dip*. See BARRIER BASIN.

basin irrigation A form of irrigation in which an earth ridge is erected around a flat area of land and water introduced to form a pond.

basin lister A modification of the LISTER PLOUGH with an attachment which forms earth ridges across the lister furrows every 4·6 to 7·6 m (15 to 25 ft), thus forming large shallow water basins.

basin recharge The portion of precipitation that remains in a basin as groundwater, surface storage, soil moisture, etc.; the difference between precipitation and runoff and other losses for a given period or storm.

basin, river See CATCHMENT AREA, RIVER BASIN.

Bateman, John Frederic La Trobe (1810–89) A distinguished British water engineer. See HAWKSLEY, THOMAS.

bat faggot A FAGGOT made from hazel, chestnut, or oak sticks without the branches. The bundles are about 0·8 m (30 in) in girth and 1·5 m (5 ft) long. See FASCINES.

bathotonic reagent A substance which diminishes the SURFACE TENSION of a fluid.

battered-bank system A heavy timber method of bank protection, similar to the RAW-BANK SYSTEM except that the bank is sloped to a convenient batter. Used in Australia and New Zealand. See also ANCHORED TREES.

battering The artificial sloping back of a river or other bank to a flatter grade to reduce caving or collapse of the earth. The procedure is often followed by BEACHING with stone or by grassing or other vegetation to bind the soil layers.

batter level A CLINOMETER for measuring the slope of BATTERS.

batters (or **side slopes**) The artificial, uniform slopes of the sides of earth dams, canals, embankments, etc. Earth embankments or farm dams should not be steeper than 1 in 3 on the water or upstream side and 1 in 2 on the downstream side (Figure B.2).

battery of wells A number of wells within a convenient radius and connected to a main pump or other lifting appliance for withdrawal of water. See WELL FIELD.

bay A portion of the sea which extends into an open-mouthed land formation.

bay (irrigation) The area of ground between adjacent CONTOUR DITCHES, CONTOUR BANKS, BORDER CHECKS, or BORDER DITCHES.

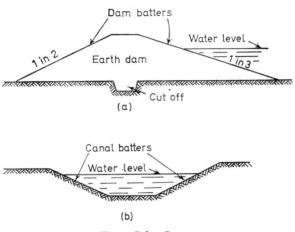

Figure B.2. Batters

bay-head bars Banks of silt or gravel formed at the heads of bays. See MID-BAY BARS.

bay-head beaches A feature associated with irregular coasts; it consists of the accumulation of water-washed sediments at the heads of bays; the material was worn and transported from the neighbouring headlands.

bay-mouth bars SPITS that have continued to accumulate and finally closed the entrance to a bay.

bay outlet or stop (irrigation) A small gated structure placed in the bank of a farm supply ditch to allow water to pass from the ditch on to the irrigation bay. The outlet is usually rectangular and made of concrete or steel.

Bazin's formula Empirical formula proposed by Bazin in 1897 giving the value of the coefficient C, in CHEZY'S FORMULA, in terms of hydraulic mean depth R and a coefficient of roughness M, a numerical constant depending

20

on degree or kind of roughness. Thus:

<table>
<tr><td>Metric units</td><td>Imperial units</td></tr>
</table>

$$C = \frac{87}{1+M/\sqrt{R}} \qquad C = \frac{157\cdot6}{1+M/\sqrt{R}}$$

beached bank The part of a river bank covered with stone above and below the normal water level. See BEACHING.

beaches The shore areas of the sea or lake washed by the tide; all shoreline deposits consisting of sand, pebbles, or boulders, and formed principally by the action of waves and long-shore currents. In general, beach sediments are coarser at the top than they are near the water, and along sea beaches pebbles often accumulate above high-tide level, grading downward into sand and finally mud below low-tide level. Mud flats are usually exposed at low tide along very low shelving shores.

beaching A layer of stones for revetting below the level of stone pitching a reservoir or embankment: the layer is from 0·3 to 0·6 m (1 to 2 ft) thick and stone size from 75 to 200 mm (3 to 8 in). Also applied to a layer

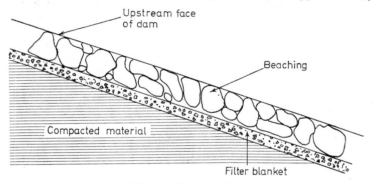

Figure B.3.   Beaching

of stones of about 150 mm (6 in) diam. to prevent scour due to wave action (Figure B.3). See STONE RIPRAP.

beach ridge A deposit of beach sediment formed by wave action. Frequently formed along a relatively flat beach where the UNDERTOW is weaker than the inward movement of the waves, with the result that silt is shifted and cast up near the water's edge to form a beach ridge. See BARRIER BEACH.

beam engine A vertical type of steam engine used widely in the early days of mining to drive the CORNISH PUMP (q.v.).

bear-trap dam A type of dam or obstruction consisting of hinged leaves. The pressure of water admitted to the inside causes the leaves to be raised and held up; the obstruction is lowered by draining the interior.

Beaufort scale   An arbitrary scale, with a range from 1 to 10, for indicating wind velocities and based on observed effects, such as bending of trees.

bed building discharge   See DOMINANT FORMATIVE DISCHARGE.

bed-building stage   The stage of a stream when there is maximum silting or building up of the bed by deposition of sediment. See ALLUVIATION, OLD RIVER, SILTING.

bedding, land   Elevation of surface of fields into a series of parallel beds or 'lands' by grading or ploughing and separating them with shallow surface drains.

bed erosion   The deepening of a stream by erosion of its bed. It may be caused by artificial means, such as bend cutting, CANALISATION, or GROYNES. Bed erosion increases channel capacity but may create other problems if excessive. See TRAINING WALL, CHECKDAMS.

Bedford Level (or Fens)   An area of land about 112 km (70 miles) long and 32 to 64 km (20 to 40 miles) broad, embracing parts of Cambridgeshire, Lincolnshire, Northamptonshire, Suffolk, and Norfolk (U.K.). The area was drained and reclaimed by the Earl of Bedford and the Dutch engineer Cornelius Vermuyden in the 17th century. The Fens now support a larger population than most agricultural areas. See LAND ACCRETION.

bed load   The weight or volume of boulders, pebbles, and gravel rolled or moved by a stream along its bed in unit time. See LOAD.

bed load function   Relationship between bed load and discharge in a given cross section of stream.

bed load sampler   An appliance for measuring the rate of movement of material along the bed of a stream; the thickness of material differs at different periods and for different streams.

bed profile   A curve indicating the elevation and shape of the bed of a stream. It may be a longitudinal curve or a transverse curve at a cross section. See STREAM PROFILE.

bed, river   See RIVER BED.

bedrock (or rockhead)   The plane or division between the surface UNCONSOLIDATED MATERIAL and the solid rocks; the first hard, solid rock encountered in a well, borehole, or excavation.

bed scour   The wearing away of the rock, soil, or deposited material from the bed of a river or sea by flowing water or waves. See LOAD, SCOUR, SCOUR, PROTECTION.

bed slope   The inclination of the bed of a stream along its course, given as difference in elevation per unit horizontal distance.

beheaded stream   A stream which has been severed from its original drainage area by another and more powerful stream; the area now draining into the second stream. See RIVER CAPTURE, OXBOW LAKE.

bela   A term commonly used in India for a sandbank or island in a river. See BAR, SHOAL.

Belanger's critical velocity   See CRITICAL VELOCITY.

bell buoy A buoy which gives audible warning by an attached bell. See CAN BUOY.

bell dolphin (or **Baker bell dolphin**) A large bell-shaped steel or concrete fender for mooring ships at sea. The bell is mounted on a cluster of driven piles of timber, steel or concrete; first used at Heysham jetty. See DOLPHIN.

bellmouth overflow A structure built on a reservoir to pass overflow discharges; consists of a vertical tower with a bellmouth inlet which passes the water downwards to an outlet conduit. Two bellmouth overflow and shafts may be constructed, one near each end of the dam.

belt, meander See MEANDER BELT.

belt of phreatic fluctuation The mass of rocks in the LITHOSPHERE in which FLUCTUATION OF WATER-TABLE occurs.

belt of soil water The upper near-surface part of the ZONE OF AERATION containing SOIL MOISTURE.

belt of weathering The belt of ground extending from the surface down to the comparatively slight depths where the weathering processes become inactive.

bench flume An open conduit which rests on a bench cut and usually constructed of reinforced concrete; frequently adopted when the soil materials are unsuitable for earth channels (Figure B.4).

bench mark (or **B.M.**) A fixed point or mark the level of which relative to a given datum is known; used by engineers and surveyors to determine the

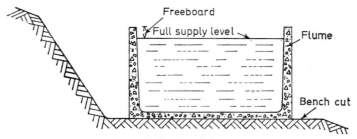

Figure B.4. Bench flume

level of other points or surfaces such as reservoirs, dams, canals, streams, and harbours. See CONTOUR LINE, ORDNANCE DATUM.

bench of silt A bed, bank, terrace, or flat of silt material bordering a river or other body of water. It may form banks or flats along the lower reaches of a river or in the channel proper where flow is sluggish. See RIVER FLATS, TERRACES.

bench terrace A terrace or small embankment of earth formed along the contour of sloping ground to regulate runoff and reduce erosion. A bench

23

terrace has a level or nearly level top and a steep or vertical downhill face. See RIDGE TERRACE.

benefit ratio The ratio of estimated benefits to estimated cost of any drainage or river improvement scheme. In a small drainage scheme the benefits may be increased productivity of the land and assessed by results of similar works in the area. The benefits of a large river improvement scheme are usually more difficult to estimate and require a study of costs in conjunction with the estimated gain in productivity of the land protected or reclaimed from flooding or overflows. See LAND ACCRETION.

bentonite A very fine clay substance with a maximum particle size in the 2 micron range; has a very high swelling property when moistened. Used in reservoirs, ponds, irrigation and water supply channels, and in small farm dams to reduce seepage. Typical application is about 5 kg/m² (1 lb/ft²) of surface. The bentonite is covered by a SAND BLANKET to prevent cracking during the period when the dam or channel is empty. To support a large head of water, bentonite must be used together with sand or soil to increase its structural strength. See LINING (HYDRAULICS).

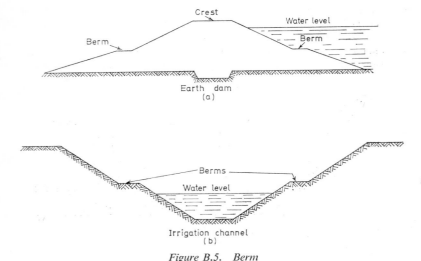

Figure B.5. Berm

berm (or **berme**) A narrow ledge; a flat shelf which breaks or modifies the uniformity of an earth slope; it intercepts earth or stones rolling down the slope and it also strengthens the embankment; often used in large earth irrigation channels and dams (Figure B.5).

Bernoulli's theorem This relates to flows in conduits and states that when a perfect incompressible fluid flows in a steady stream, then, if friction and eddy current losses are neglected, the total energy is constant. The equation is:

$$p/w + z + v^2/2g = \text{constant}$$

where p is the pressure of fluid, w = specific weight of fluid, z = height of fluid above an arbitrary datum, v = velocity of fluid, and g = acceleration due to gravity.

berth Place where a ship lies at anchor, or at a wharf for loading or unloading.

berthing impact The forces to which structures, such as piers and jetties, are subjected during the berthing of vessels; generally estimated from the kinetic energy of a large vessel and an assumed berthing velocity of about 150 mm/s (6 in/s).

betrunked Applied to the partial submergence of a river system; streams that were formerly connected flow into the sea by separate mouths. See BEHEADED STREAM.

bifurcation structure A structure or gate placed in a conduit to divide the water flow into two separate conduits.

billabong (Australia) See OXBOW LAKE.

biological filter See BACTERIA BED.

blanket grouting A form of grout treatment used under concrete dams and to a lesser extent under earth dams. It is used to consolidate the foundations of a dam where the foundation rock is of good quality but jointed. Usually the entire foundation area to a depth between 6 and 15 m (20 and 50 ft) is treated with low-pressure grouting through holes drilled on a 3 or 6 m (10 or 20 ft) grid. The grout pressures are controlled to prevent the upheaval of the foundation rock. See GROUTING.

blind drain See RUBBLE DRAIN.

blizzard A violent snowstorm with intense cold and high wind.

block-in-course Blocks of hard stone, of variable length but uniform depth not exceeding 0·3 m (12 in) laid in courses in dock walls.

blockwork Stone or precast concrete blocks weighing from 10 to 50 t used in the construction of BREAKWATERS and similar marine structures to resist wave action. See RUBBLE-MOUND BREAKWATER, TETRAPOD, TRIBAR.

blowing well A water well from which a current of air is periodically blown out. See BREATHING WELL.

blow-off An outlet fitted to a pipeline for emptying a low sewer or for discharging water or sediment. See AIR VALVE.

blow-out (1) A failure due to TUNNELLING or PIPING in earth dams or channels. (2) The sudden loss of compressed air from a tunnel or caisson, very dangerous in certain conditions.

blow-well See ARTESIAN SPRING.

bluffs High banks presenting steep fronts, as at the extremes of river meanders.

B.M. See BENCH MARK.

25

body current (lakes)   The general movement of the water from inlet to outlet of a lake; the velocity is low and may be noticeable only at the head or foot of the lake. The flow is not necessarily direct from inlet to outlet. See SEICHES.

boil   (1) The seepage or inflow of water and fine sand or silt at the bottom of an excavation due to a high external water pressure. The remedies include reducing the pressure difference or drainage or lowering the water-table in the vicinity of the excavation. (See PIPING, WELLPOINT.) (2) A vertical eddy developed in a river during a flood. It produces an upthrow of water to the surface, causing it to 'boil'. See also SAND BOIL, TIDAL CURRENT.

boison   A topographic basin where the streams converge to a central area or point. See RADIAL DRAINAGE.

bollard   A strong cast-iron post fixed in a quay structure for mooring vessels. See BELL DOLPHIN.

bonded gravel screen   A WELL SCREEN of an improved type and made from particles of gravel bonded together with a resin adhesive. Its yield performance is at least equal to the ordinary metal well screen but it is much cheaper to construct. It was developed by scientists of the State Rivers and Water Supply Commission, of Victoria, Australia.

bone oil (or **Dippel's oil**)   A polluting substance sometimes used by military forces to make water supplies undrinkable. When added to water it gives a nauseating smell and a taste so objectionable that it cannot be drunk in sufficient quantities to poison a person, hence it does not contravene the Hague Convention.

boning   A field method for obtaining levels or of ensuring a correct gradient of pipelines with the aid of BONING RODS.

boning rod   A 'T'-shaped staff of timber about 1·2 m (4 ft) long and rods used to set individual pipes on the correct gradient, or to extend given points at the same level. See SIGHT RAIL.

boom   An obstruction or a floating barrier of timber extending across a harbour or river mouth.

booster pump   An automatically controlled variable-speed impeller-type pump with electric drive is often favoured; usually installed on the track of the main to increase the pressure. A booster pump is sometimes used to feed small areas situated above the level of the source in a gravity system.

border check   A ridge or mound of earth formed down the slope of an irrigation area to control the spread of water.

border check irrigation   The use of low earth banks to confine the water to runs or bays; often adopted in pasture irrigation. See CONTOUR FURROWS.

border dikes (U.S.A.)   Earth ridges erected around a field or area to impound irrigation water. See DIKE.

border ditch   A shallow channel or ditch formed down the slope of irrigated land. The ditch facilitates drainage and the ditch banks control the spread of water.

26

border irrigation Flood irrigation of land areas between BORDER DIKES.

bore See TIDAL CURRENT. Also applied to internal diameter of a pipe.

borehole A hole drilled into the ground, from the surface or from subsurface excavations, to secure geological information or for the drainage or abstraction of water, or for access to hydraulic works, etc. In the U.K. the trend is towards larger boreholes, from 0·6 to 1·0 m (24 to 40 in) and depths for tapping water from 120 to 180 m (400 to 600 ft) and they rarely exceed 300 m (1000 ft). See WELL.

borehole casing A plain or perforated pipe of steel or other material inserted in a borehole; often used in weak or loose ground. The pipe sections may be coupled or flush-jointed. The latter is smooth externally and internally. In recent years, bronze, stainless steel, asbestos cement, and plastic casings have been used. See OPEN-END WELL, PERFORATED-CASING WELL, SCREENED WELL.

borehole log A record, mainly of the rocks penetrated during the drilling of a borehole, and prepared by the master driller or field geologist. The information recorded varies with the objectives in putting down the hole. See WELL RECORD, WELL SECTION.

borehole pattern The disposition or spacings of a group of boreholes put down for water supplies or for geological information. Where the deposits are relatively uniform, the bores may be positioned at the corners of squares or equilateral triangles. The pattern depends on the objectives and the economic aspects.

borehole pump In general, a CENTRIFUGAL PUMP, electrically driven and designed as a narrow vertical chamber. It may be used to provide water for farming, for irrigation, or for dewatering areas. See SINKING PUMP, SUBMERSIBLE PUMP.

borehole samples Refers to the soil or rock chippings, sludge, or core extracted during drilling and used to ascertain the nature of the ground penetrated in the borehole. The diamond drill and shot-drill yield rock cores, while percussive drills yield chippings and sludge. Samples can also be obtained by earth augers. See CORE, SLUDGE SAMPLE.

borehole surveying The lowering of instruments into a borehole to determine whether the hole has deviated from the vertical plane or the planned direction of drilling. Horizontal or angle diamond drill holes over about 90 m (300 ft) deep are liable to deviate badly. Vertical holes may also become crooked, but not usually to the same extent. Deviations may be caused by (1) lack of care or drilling defects, (2) shear zones or fault planes, or (3) inclined rocks of varying hardness.

bore, hydraulic See HYDRAULIC BORE.

bore, tidal See TIDAL BORE.

borrow pit An excavation or area from which soil or rock is taken to build an earth dam, embankment, or other construction (Figure F.1). See CONSTRUCTION MATERIALS.

bottle silt sampler   A device for taking samples of silt and consisting of a bottle held in a frame and suspended from a pipe. A lever at the top of the pipe operates a rubber stopper fitted to the bottle. See BOTTOM SAMPLER.

bottom contraction   A reduction in cross-sectional area of overflow water caused by contraction of nappe by crest of weir. See END CONTRACTION.

bottom ice   See SUBSURFACE ICE.

bottomland   A U.S.A. term for FLOODPLAIN.

bottom sampler   An appliance for picking up sample material from the bed of the sea or on other body of water; consists of a SOUNDING LEAD with adhesive on the underside. See BOTTOM SILT SAMPLER.

bottomset beds   See DELTA.

boulder   A large block of rock which has been transported some distance from its parent bed by ice or water and become rounded.

boulder clay   A clayey deposit containing boulders which often show striations indicating glacier movement; the deposit is typically unstratified. See GLACIAL DRIFT.

boulder well (or **cavity well**)   A water well put down into a thick aquifer composed of sand, gravel, and boulders.

bound gravel   Hard rounded or lenticular masses of gravel and sand some times occurring along the zone of fluctuation of the water-table; formed by the deposition of cementing substances around the constituent grains. Bound gravel may cause difficulties when putting down water wells; also, the masses may be mistaken for bedrock.

Bourdon pressure gauge   An instrument which can be used to measure water pressure and PORE WATER PRESSURE. It consists essentially of a tube, of oval cross-section, which tends to straighten as the internal pressure is increased.

bournes   VALLEY SPRINGS in which the flow is usually confined to winter rainfall, during which period they give rise to streams known as LAVANTS. BOURNES or NAIL-BOURNES are terms also applied in some districts (the Chalk of the U.K.) to temporary streams formed in otherwise DRY VALLEYS after heavy rainfall or very wet seasons.

box caisson (or **American caisson** or **stranded caisson**)   A large box of steel, or usually reinforced concrete, with an open top, constructed on shore and floated out into a river or seaway and sunk at the site selected for a foundation to form part of the permanent structure. The box, which enables excavation to be done in relatively dry conditions, is used for bridge piers and similar works. See CAISSON.

box culvert   A box-shaped conduit for carrying a small flood flow under roads or embankments. See CULVERT.

box dam   A COFFERDAM which entirely encloses a site or area.

box drain   A small box-shaped brick or concrete drain.

box gauge   A tide GAUGE operated by a float inside a vertical box with openings in the base for the entry of tidal water.

brackish   Water which ranges from 1000 parts per million (p.p.m.) ,or

28

milligrammes per litre (mg/l) up to the dissolved salt content of sea water. *Mildly brackish:* 1000 p.p.m. to 5000 p.p.m. (mg/l). Moderately brackish: 5000 p.p.m. to 15 000 p.p.m. (mg/l). *Heavily brackish:* 15 000 p.p.m. to 35 000 p.p.m. (mg/l). *Sea water* generally contains 35 000 p.p.m. (mg/l). (U.S.A. Bureau of Reclamation's 1966 Annual Report of Progress of Engineering Research.) See BRINE, FRESH.

Bradford gauge A large-capacity RAIN GAUGE used for weekly or monthly readings at less accessible stations, in which the amount of water collected is measured by means of a graduated rod immersed in the liquid. See RAIN RECORDER.

braided course A very wide river which has separated into a number of entwined channels called ANABRANCHES. The waters usually carry a large amount of detritus.

Bransby-Williams formula A British formula to calculate the TIME OF CONCENTRATION:

$$t_c = \frac{0.88L \times 60}{M^{0.1} S^{0.2}}$$

where t_c is the time of concentration (minutes), L is the length from source to given point under investigation (miles), S is the average slope of length (percentage), and M is the area of drainage basin (square miles). See FLOOD FORMULAS.

breadth (stream) See WIDTH.

breakaway Sometimes applied to a new channel cut by a river during flood periods. See CUTOFF.

breakers (waves) Waves which collapse or whose crests fall over. Waves approaching a shelving shore change in form and become both higher and shorter; the crests become steeper and sharper and finally they collapse. Waves of a given height break in the same depth of water, and the zone or LINE OF BREAKERS is that along which the incoming waves collapse. See FORCED WAVES.

break-pressure tank A small open tank sited on a gravity pipeline to reduce the maximum pressure of water in the pipe column. A series of such tanks is used and placed at the level of the HYDRAULIC GRADIENT of a gravity water main in undulating country. Each tank in turn takes the discharge from the main and a new section of main takes off and enters the next tank in the series. Since the water main operates under a much lower pressure than that in a continuous main, considerable savings in costs are often possible.

breakwater (or **mole**) Any structure to break the force of waves; applied especially to a wall constructed out into the sea to safeguard a natural or artificial harbour. See BLOCKWORK, FLOATING HARBOUR, RUBBLE-MOUND BREAKWATER, TETRAPOD.

breathing (river) Refers to the rise and fall in water level of a river. See TIDAL RIVER.

breathing well A water well in which air is alternately blown out and sucked in; the phenomenon appears to be related to barometric pressure. The blowout of air is usually strong and noticeable while the intake of air is not so apparent. See PERIODIC SPRING.

breeze concrete A concrete with poor fire-resisting qualities; however, it is cheap and nails can be driven into it. Made by mixing 3 parts of coke breeze (the finer coke products from coke ovens), 1 of sand, and 1 of PORTLAND CEMENT. (See STANDARD MIX.)

bridge abutment That portion of a bridge works joining the roadway to the bridge structure proper. Due to the contraction of this point and consequently the increased velocity of flow, the abutment is protected by suitable anti-erosion works. See ABUTMENTS, WING WALL.

brine Any water in which the content of dissolved salts is greater than sea water, such as the Dead Sea. See FRESH.

British Rainfall Annual volume of the British Rainfall Organisation Meteorological Office, London. In the U.K. rainfall data have been systematically collected since 1860 and summarised in *British Rainfall.* See ISOHYETALS.

broad-base terrace A ridge-like terrace with gently sloping sides, a rounded top with a dish-shaped channel to control erosion by diverting runoff along the contour of sloping ground at a non-scouring velocity. The terrace is from 4·5 to 9 m (15 to 30 ft) wide and from 0·25 to 0·5 m (10 to 20 in) high and may be level or have a grade towards one or both ends. (See CONTOUR FURROWS.)

broadcrested weir A WEIR with a flat or gently sloping crest, with a length along the line of flow considerably greater than the height of nappe over it (Figure B.6). (See BARRAGE.)

broad irrigation A method of sewage disposal, without pipe drains, sometimes used on agricultural or waste land. The sewage flows over and soaks

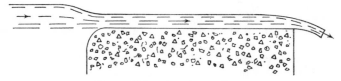

Figure B.6. Broadcrested weir

into the ground, which has been carefully levelled to avoid local accumulations of sewage.

brook A general term for a small stream flowing in a natural surface channel; the flow is usually turbulent and the channel rugged and stony with many irregularities; usually restricted to watercourses with a discharge below about 0·14 cumecs (5 cusecs).

30

brush matting A matting or cover consisting of branches placed on land to reduce erosion and conserve moisture while some form of vegetal cover is being established; a matting consisting of mesh wire and brush placed along a stream bank to reduce erosion. See RIPRAP.

brushwood Small tree branches, saplings, or shrubs often used in river improvement work. They are placed close together and parallel. The material should be strong enough to resist breaking, thin enough to be flexible, and reasonably straight. Green willow branches and the straight-growing stems of the tea-tree make good brushwood. See FASCINES.

brushwork See BRUSH MATTING, BRUSHWOOD, FASCINES.

BS British Standard.

B.S.I. BRITISH STANDARDS INSTITUTION (q.v. Appendix).

bucket (1) A cup on the perimeter of a PELTON WATER WHEEL; the container of a bucket elevator or dredger. (2) The reversed curve in a spillway profile which deflects the water horizontally at its base from the steep overflow face on to apron (U.S.A.).

bucket-ladder dredger See DREDGER, ELEVATOR DREDGER.

bucket pump A reciprocating pump sometimes used in wells or boreholes or for dewatering purposes. See LIFT PUMP.

buffer strips Strips of erosion-resisting vegetation or grass formed along the contour below or between cultivated fields.

built terrace Material deposited on the sea floor beyond the platform or in sheltered areas in bays; a beach deposit consisting of material derived from the weathering of a headland and shore platform.

bulk density The weight per unit volume of any material including moisture, generally expressed in kg/m^3 (lb/ft^3). See DRY DENSITY.

bulkhead A partition or structure in a conduit or tunnel to stop the flow of water; upright partition in a ship to form separate compartments.

bulking The increase in volume or swelling of a dry sand or shale when it absorbs water. The increase in volume when rock or soil is excavated ranges up to 50%. See SWELLING PRESSURE.

bulldozer A high-powered caterpillar-mounted tractor to the front of which is fitted a concave buffer-blade, 0·6 to 0·9 m (2 to 3 ft) high and 1·8 to 3·0 m (6 to 10 ft) long. As the machine moves forward the blade levels the ground. Used at earthworks to smooth out irregularities. See TILT DOZER, SCRAPER.

bunds A continuous vertical or near-vertical wall or structure acting as a bank protection along the front of the margin; it gives strength and stability in the space available. In India the term is applied to a LEVEE but in other parts of the Far East it refers to almost vertical structures to retain earth along quays and riverside drives.

Bunter The Bunter beds of the Triassic form important aquifers in the Midlands of the U.K. The rocks are highly permeable and yield water readily, once the boring has entered the water-table. In general, the water is of good quality and hardness is low and mainly temporary. See COAL MEASURES.

31

buoy An anchored float to show navigable courses, reefs, or natural obstructions such as bars. See ANCHOR BUOY.

buoyancy The upward thrust exerted by a fluid upon a body immersed in the fluid; equal to the weight of the fluid displaced. See ARCHIMEDES'S PRINCIPLE.

buoyant foundation (or **buoyant raft**) A foundation consisting of a continuous slab of reinforced concrete which extends under the entire structure or works; used in areas such as river estuaries where the supporting material consists of nearly fluid silt or mud. It is so designed that the weight of the concrete raft plus the load it carries is about equal to the weight of the saturated silt it displaces. See RAFT FOUNDATION.

buried channels Usually pre-Glacial depressions or valleys filled with glacial drift. Many of the present streams occupy the wholly or partly filled pre-Glacial valleys. Tunnels and other works in these channels often present engineering problems, extra costs, and perhaps modifications in design. On the other hand, the channels often yield useful supplies of shallow water.

buried erosion surface A surface of UNCONFORMITY below ground level.

bush logs Rough, unshaped tree trunks, of various lengths and sizes, for use in river and bank protection works. They are usually cut at, or near, the site where they are required for use.

butterfly gate A GATE used on hydraulic structures; similar to the BUTTERFLY VALVE, it is easily operated because it is perfectly balanced.

butterfly valve A circular disc fitted inside a pipe and hinged at two pivots on a diameter; often used in large pipes for controlling the flow, especially in HYDROELECTRIC POWER schemes. The valve opens and closes very easily because it is well balanced.

buttress A concrete or masonry projection or pier built out at right angles to a wall to reinforce its resistance to earth thrust or water pressure.

buttress dam A modified form of GRAVITY DAM in which buttresses are incorporated to direct thrusts to the foundation. A buttress dam, as compared with a gravity dam, requires a considerably greater quantity of shuttering. In the MASSIVE BUTTRESS DAM, the upstream face may be flat and nearly vertical, or each buttress may have a steeply inclined round-head or diamond-head. All the above types were used in the Scottish hydroelectric schemes. See PRESTRESSED GRAVITY DAM, ROUND-HEAD BUTTRESS DAM.

butts The thickest ends of trees cut down and removed. When used for bank protection, the butts point upstream and are well anchored. See RAW-BANK SYSTEM.

bye channel See BY-WASH.

by-pass A conduit or pipe arrangement for directing water flow around instead of through another conduit, pipe, or valve.

by-pass channel An artificial channel for directing flood water in excess of a river's capacity; the channel is often lined with LEVEES on each side to increase its capacity.

by-wash (or **bye channel** or **natural escape**) A type of spillway or escape channel for a small reservoir. It passes flood waters around a dam, as distinct from spillways passing flood water through or over a dam. It is termed a masonry or paved escape when such protective works are required, or a NATURAL ESCAPE when the ground is not paved. Some engineers consider that a by-wash for large FARM DAMS should be capable of handling flows for a 1 in 100 year flood. The failure of many farm dams is probably due to a faulty, or no, by-wash (Figure C.1). See DROP-INLET SPILLWAY, OVERSHOT SPILLWAY.

C

cable tool A sharp chisel-edged bit used in drilling a deep well; it is lifted and dropped to break the rock by impact. The rock fragments or cuttings are removed from the hole by a bailer, which is a section of pipe with a foot-valve through which the cuttings enter. See ROTARY DRILL.

caisson A rectangular or cylindrical watertight structure for holding back water or waterlogged sediments from a site while digging proceeds to solid ground upon which a foundation can be constructed. The caisson finally forms part of the foundation structure. See OPEN CAISSON, SHIP CAISSON.

caldera lakes Lakes that occupy enlarged volcanic craters. They are uncommon and the water is of little value.

Calgon method The use of Calgon (sodium hexametaphosphate) to increase the supply of water from boreholes in chalk, clays, and silts. See SWABBING.

calm belt See DOLDRUMS.

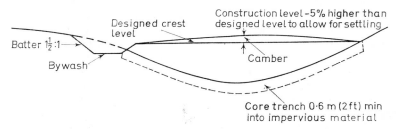

Figure C.1.   Camber

camber (1) The additional bulk or height given to a dam embankment to allow for settlement of the material (Figure C.1). (2) A small tidal basin or dock.

camel A large hollow steel float secured to a ship to elevate and float it across a shallow area. See SAUCER.

camp sheathing (or **camp shedding** or **camp sheeting**) A wall to hold back a river bank; construction often consists of two connected rows of timber piles 1·8 to 3·0 m (6 to 10 ft) apart with earth packed into the space between. See BARGE BED.

canal An artificial waterway or channel, usually for the passage of vessels or boats. It may also be used for connecting two or more bodies of water, for irrigation or water power or other applications. See LATERAL CANAL.

canal, inundation See INUNDATION CANAL.

canalisation (1) A defined channel cut through a swamp area with levee banks on both sides which are constructed from material excavated from the channel. The canalised stream takes the floodwater from the upper reaches through the swampy ground, while the rain falling on the swamp itself may be removed by independent drains. The canal capacity is often designed to cope with small to medium floods. Drop structures may be necessary to reduce the grade and to give a non-scouring velocity. (2) The division of a river into reaches separated by dams, weirs, and control structures. Canalisation assists navigation, prevents or reduces flooding, and provides water for irrigation and sometimes HYDROELECTRIC SCHEMES.

canal lift A mobile tank to allow barges to pass through a lock with a lift exceeding about 15 m (50 ft). The tank is hauled on wheels up or down an incline, or lifted and lowered vertically. See LIFT.

canal, main irrigation (or **feed canal**, or **main canal** or **storage canal**) A canal conveying water from a stream to an off-channel storage reservoir; a conduit or channel conveying water from a storage reservoir or point of diversion to lateral or branch canals or to a group of irrigated farms.

canoe fold See SYNCLINAL CLOSURE.

canyon A deep narrow valley or gorge, often with a swiftly flowing stream. Arid climate and firm rocks are conditions favourable to the development of canyons. The Colorado canyon is a good example.

capacity Receiving power or holding power; ability (as wells) to yield water or streams to transport sediment; volume of artificial or natural containers, storage, or basins, etc. See CAPACITY OF STREAM, CAPACITY OF WELL, GROSS CAPACITY, INFILTRATION CAPACITY, INVERTED CAPACITY, MAXIMUM CAPACITY, OUTLET CAPACITY, SPECIFIC CAPACITY, TESTED CAPACITY, TOTAL CAPACITY, WATER CAPACITY, WATER-HOLDING CAPACITY.

capacity curve (or **storage curve** or **volume curve**) (1) A graph showing the relation between the volume of water in a tank, reservoir, or storage and the level of the water surface (Figure C.2). (2) A graph of the capacity rate of flow in a conduit, or pipe.

capacity factor In the case of channels or canals for irrigation, the ratio of average supply to authorised full supply or capacity.

capacity formula A hydraulic formula to determine the discharge or capacity of a channel:

$$Q = AV$$

where Q is the discharge or capacity (cumecs or cusecs), A is the cross-

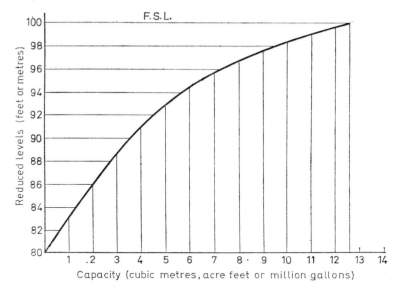

Figure C.2. Capacity curve

sectional area of channel (m^2 or ft^2), and V is the mean velocity of water (m/s or ft/s).

capacity of stream (1) The ability of a stream to transport material in terms of total weight. See LOAD, TRANSPORT COMPETENCY. (2) The largest flow of water a stream can carry without overflowing its banks.

capacity of well The ability or competence of a well to yield water under certain specified conditions. See ECONOMIC YIELD, MAXIMUM CAPACITY, TESTED CAPACITY (OF WELL), TOTAL CAPACITY.

capacity of well, maximum See MAXIMUM CAPACITY (OF WELL).

capacity of well, tested See TESTED CAPACITY (OF WELL).

capillary action A generic term for the movement of liquids due to capillary forces, or forces due to unblanced inter-molecular attraction at the liquid boundary; the rise or movement of water in the interstices of a soil or rock specifically in the CAPILLARY FRINGE.

capillary fringe The zone or belt of ground immediately above the water-table and containing FRINGE WATER or CAPILLARY WATER.

capillary porosity See TOTAL POROSITY.

capillary pressure (or seepage force) If the ground is being drained from inside an excavation, the capillary pressure tends to cause instability and the earth faces may collapse. If the ground is being drained at points outside an excavation (see WELLPOINTS) the capillary pressure may help to stabilise the earth. The capillary pressure varies with the nature of the

35

ground and ranges from 0·0063 to 0·063 N/mm² (0·9 to 9 lbf/in²) in silt and much lower in sand.

capillary rise (or **height of capillary rise**) The height to which water rises above the free water level owing to CAPILLARY ACTION.

capillary water Water retained in rocks or soils above the water-table; water in the CAPILLARY FRINGE.

captive float (or **reefing float**) A canvas skirt, with stiffening rings spaced at regular intervals, supported by a float resting on the surface of water. By reefing, or clamping the required number of stiffening rings to the float to shorten the canvas skirt, its length is adjusted to the depth of the channel.

capture, river See RIVER CAPTURE.

carbonate hardness A term which has replaced the earlier TEMPORARY HARDNESS. See HARDNESS OF WATER.

Carboniferous Limestone The Carboniferous Limestone in the U.K. has abundant water occupying fissures which also feed large springs. Bristol takes the water from the Cheddar Springs. In parts, much of the water flows along underground streams which form useful supplies when tapped. The water is very hard but usually clear and palatable; surface pollution is a hazard. See WEALDEN.

carry-over storage (1) A reserve of water retained in a reservoir after wet or normal years to ensure that the minimum outflow demands are met during drought years. (2) A reservoir which meets these requirements.

cascade (1) A stream irregularity intermediate between a WATERFALL and a RAPID; a stretch of stream where the slope is considerable but not sufficient to cause the water to drop vertically. (2) A series of vertical steps in a sewer or drain, separating lengths where the slope is normal. A cascade avoids blockages with small or sluggish flows or scouring of pipe by high flows on steep slopes.

cataract A rush of water; a downpour of rain, or, especially, the flow of a large body of water over a precipice. The most famous falls of water are the Victoria Falls on the Zambesi in Africa, the Niagara in North America, the Orinoco in South America, the Falls of the Rhine at Schaffhausen, and the Cascade of Gavanni in the Pyrenees. See WATERFALLS.

catastrophic floods See FLOOD (return period).

catch basin A CATCH PIT (U.S.A.).

catch drain See CATCHWATER DRAIN.

catch feeder A DITCH used for irrigation.

catchment area (or **drainage area** or **drainage basin** or **gathering ground**) The area of land from which rainwater or snow melt drains into a reservoir, pond, lake, or stream; usually expressed in hectares or square kilometres (acres or square miles). Catchment areas are often managed to slow down runoff, absorb precipitation, and avoid erosion.

catchment area of aquifer See INTAKE AREA OF AQUIFER.

catchment management Management of catchments to improve the quality of runoff water. The measures include safety against fire and damage, prevention of erosion, and maintenance of ground cover. In general, the forested catchment gives the best results. See FIRE CONTROL (CATCHMENTS).

catchment yield The total volume of water which flows from a CATCHMENT AREA over a given period; often expressed as the equivalent depth of water in mm (inches) over the catchment area.

catch pit A pit or small chamber located at an accessible point in a drain, ditch, or sewer to collect solid matter and thus prevent blockage of the inaccessible parts of the drains. The pit is periodically cleaned.

catchwater drain (or **catch drain** or **catchwater** or **drain** or **interception channel**) A surface ditch or drain on the rise side of an excavation or earthworks to intercept and divert surface water clear of the working site or urban water storage. See GRIP.

catchwaters Open channels or pipes to intercept the runoffs from adjoining catchment areas and conduct the water into the IMPOUNDING RESERVOIR. The channels are formed at steady slopes gradually crossing the contours. Pipes are often preferred, although both methods have their merits and demerits.

caterpillar gate Used in U.S.A. for a heavy steel gate in a spillway for controlling the flow. It is mounted on crawler tracks with steel rollers and operates on inclined rails along each side of the opening.

causeway A raised road or way over water or wet marshy ground; structure may consist of an earth bank or fill material and a paved surface.

caverns Solution channels and cavities formed in rocks, usually limestone, by water percolating along joints, bedding planes, and other paths of weakness. See SINKS, SOLUTION CAVITIES, SWALLOW HOLES.

cavern water Water occupying caverns and solution cavities. See SUB-TERRANEAN STREAM.

caving The wearing away, undercutting, and collapse of soils and rocks forming the banks of a river by the action of flowing water. See UNDERCUT SLOPES, UNDERCUTTING.

cavitation The formation of cavities during pumping at high speeds, resulting in corrosion of metal parts due to liberation of oxygen from the water. It may occur with centrifugal pumps or in water turbines near the DRAFT TUBE.

cavity well See BOULDER WELL.

celerity of wave The velocity of propagation of a wave through a liquid, compared with the undisturbed velocity of liquid through which the disturbance is propagated. The celerity of a gravity wave of relatively large wavelengths in an open channel is equal to $\sqrt{(gd)}$, where d is depth of water and g is acceleration due to gravity—9·8 m/s^2 (32·2 ft/s^2).

cellular cofferdam A COFFERDAM with a double wall consisting of a succession of cells in contact. Each cell may consist of a steel SHEET PILE ring, diam. 18·3 m (60 ft) and filled with sand. This is an American DOUBLE-WALL COFFERDAM, sometimes adopted in major schemes.

37

cementation   Processes involving the injection of cement grout into fissured rocks to render them relatively watertight during excavation work, tunnelling, and shaft sinking; also used extensively in earth and concrete dams and cutoff works for the prevention of leakage. See GROUTING.

cement ditches   See LINED DITCHES.

central drainage tunnel   See DRAINAGE TUNNEL.

centre of pressure   The centre of pressure of a plane immersed surface is the point in the surface through which the resultant of all the pressure on the surface acts. The principle is applied in SELF-ACTING MOVABLE FLOOD DAMS.

centrifugal pump   A pump in which the water enters near the centre of a high-speed rotating IMPELLER and through which it flows radially under centrifugal force. Its kinetic energy is then converted into pressure energy in the casing. It is a compact pump and suitable for direct electric drive, but the water must be reasonably clean. Centrifugal pumps of various types are widely used for irrigation work and water engineering generally. See PUMP.

centrifuge   A machine rotating at a very high speed and used widely for cleaning, sizing, and dewatering minerals; also used in soil mechanics and in sewage for dewatering sludge.

centrifuge moisture equivalent (or **CME**)   The percentage of moisture retained by a soil sample after being first saturated with water and then subjected to a force equal to 1000 times the force of gravity, usually for $\frac{1}{2}$ hour or 1 hour. Sometimes used in U.S.A. for comparing road soils.

centroclinal dip   A condition where the rocks or ground dip towards a common point or area. See BASIN.

cess   A drain pipe, often porous, laid along the foot of channel excavations. It may be 230 mm (9 in) or more in diameter.

cesspit (or **cesspool**)   A brick- or concrete-lined tank placed underground for collecting sewage in districts where no sewage treatment exists. The pit is pumped out periodically. The cesspit is becoming obsolete and the SEPTIC TANK is preferred.

Chadwick, Edwin (1800–90)   A pioneer in the development of sanitation in the U.K.; largely responsible for the passing of the first Public Health Act in 1848.

chain gauge   A gauge consisting of tagged or indexed chain tape or other line, with a weight at the end which is lowered to touch the surface of water. Often used for water level measurements at bridges. See ELECTRICAL TAPE GAUGE, WIRE WEIGHT GAUGE.

chain of buckets   An arrangement used in the early days of mining and engineering for dewatering shaft sinkings and excavations. It consisted of a series of buckets attached to a chain or belt which was revolved around a wheel by men or animals. It was a primitive form of the BUCKET-LADDER DREDGER for removing water.

chain of locks A series of connected LOCKS in which each CHAMBER is followed immediately by another; the HEAD-GATE of each lower lock forms the tail-gate for the lock above it.

chalk The Chalk of south-east England is probably the best known and largest aquifer. The rock is relatively impermeable but water movement occurs along fissures and wells put down in fissured zones give high yields. The water is usually hard (temporary hardness). See BUNTER.

chamber (or lock bay) The space enclosed between the upper and lower gates with reference to a canal lock.

Chamier's formula A British flood intensity formula:

$$Q = 640iKa^{\frac{3}{4}}$$

where Q is the maximum flood intensity in cusecs, i is the average rate of rainfall anticipated in inches per hour for such duration as will allow of the flood water to flow to the outlet from the furthest point of the catchment, K is the coefficient of runoff, and a is the area of the catchment in square miles. (*Proc. Instn. Civil Engrs*, **134**, 319). See FLOOD FORMULAS.

channel A natural or artificial groove, ditch, or conduit for the flow of water; a WATERCOURSE or CANAL; a narrow sea; that stretch of a stream where water normally flows. See BY-PASS CHANNEL, BY-WASH.

channel axis See AXIS OF CHANNEL.

channel capacity In artificial channels, the maximum flow carried when the channel is running at FULL SUPPLY LEVEL. In natural channels, the safe maximum flow of water that a stream can carry without overflowing its banks.

channel check A small structure, with gates, built across a farm supply ditch or channel to regulate and control the flow of water, usually made of steel or concrete. See CHECK.

channel, diversion See DIVERSION.

channel improvement Mainly concerned with methods to improve the flow characteristics of a natural or artificial channel and thus increase its carrying capacity. The methods include clearing, cleaning, and excavation or dredging to enlarge the area. Sometimes applied to measures to prevent channel erosion. See RIVER TRAINING.

channel line The path or line of maximum flow along a stream; usually coincides with the THALWEG.

channel of approach A channel which leads up to a measuring weir, gate, tube, or orifice. See APPROACH VELOCITY.

channel precipitation That part of precipitation that falls directly on the water surface of a canal, stream, or any open channel.

channel revetment A lining or facing to protect a channel bank from erosion; it may consist of stone, concrete, piling, etc.

channel roughness The unevenness, irregularity, or texture of the channel surface in contact with the flowing water. See CHEZY'S FORMULA, KUTTER'S FORMULA, MANNING'S FORMULA, ROUGHNESS COEFFICIENT.

channels, pilot See PILOT CHANNELS.

channel storage The water stored in a channel during periods when the flow exceeds the discharge capacity of the channel. See BANK STORAGE.

charge, sediment See SEDIMENT CHARGE.

check A structure built to regulate or raise the water level in a supply channel; an area of land irrigated by water confined by earth ridges.

check and drop A structure which combines the functions of both a CHECK and a DROP; the water level may be raised upstream of the structure and dropped on the downstream side.

checkdam A low, temporary type of dam or barrier erected across a narrow watercourse to retard flow, minimise erosion, and promote the deposition of silt. It may be of stonemesh, with an apron of similar construction or of timber. The establishment of a thick growth of trees or bushes along the banks and bed of the channel also retards velocity and scour.

check irrigation Irrigation method whereby earth ridges are built around an area, partly or wholly, to retain water. See BASIN IRRIGATION, BORDER DIKES, SQUARE CHECK IRRIGATION.

check valve See CLACK VALVE.

Cheddar Springs See CARBONIFEROUS LIMESTONE.

chemical consolidation A chemical GROUTING process for sealing and strengthening loose heavily watered ground around excavations. A gel-forming chemical is injected into the ground and the time delay in gel formation can be controlled. A proprietary process using A.M.9. has given promising results. Chemical grouting is an improvement on methods in which it is necessary to inject two separate chemicals (calcium chloride and sodium silicate) and consolidation takes place in the ground. See JOOSTEN PROCESS.

chemical gauging (or chemi-hydrometry) A method of measuring quantity or flow of water by introducing a chemical solution of known concentration at a cross section upstream and then determining the degree of dilution of the solution at another cross section downstream. The method is sometimes used along mountain streams where current meters cannot be used and where measuring structures would be costly or difficult to erect. See DILUTION METHODS, SALT VELOCITY METHOD.

chemical precipitation The settlement of sewage aided by FLOCCULATION.

chemical weathering The action of rainwater containing weak acids on stone structures, rocks, and soils. The acids are derived from atmospheric gases such as carbon dioxide and sulphur dioxide. In industrial areas, chemical weathering may be serious on account of the higher content of corrosive impurities in the air and rainwater. The limestone or calcium carbonate is dissolved by rainwater containing carbon dioxide and is held in solution as calcium bicarbonate, thus:

$$CaCO_3 + H_2O + CO_2 \rightarrow Ca(HCO_3)_2$$

chemical water treatment The addition of chemical substances to hard water which break down the impurities. The residue is either precipitated for subsequent retention in a filter or driven off as a gas, or passed on in solution in harmless or less harmful form. Lime or soda or a combination of both is generally used with or without the addition of COLLOIDS or ZEOLITES. See COLLOIDAL WATER TREATMENT, HARDNESS OF WATER.

chemi-hydrometry See CHEMICAL GAUGING.

chemise A wall built as a protective lining to an earth bank.

chevron drain (or **herringbone drain**) A system of rubble-filled trenches or open-jointed pipes set out in herringbone pattern; the water flows into main drains laid out in the direction of maximum slope (Figure C.3).

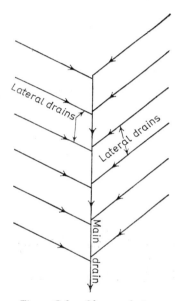

Figure C.3. Chevron drain

Chézy formula An empirical formula expressing the relation between mean velocity of flow V, hydraulic mean depth R, hydraulic gradient or slope S, and coefficient C. Thus:

$$V = C\sqrt{(RS)}$$

See BAZIN'S FORMULA, KUTTER'S FORMULA, MANNING'S FORMULA.

chimney A conspicuous pillar-like projection of rock on a coast which has been separated from the sea cliff by wave action around it. See STACK.

41

chimney drain One type of DOWNSTREAM DRAIN consisting of a vertical gravel drain varying in thickness from about 1·5 to 3 m (5 to 10 ft) (Figure C.4).

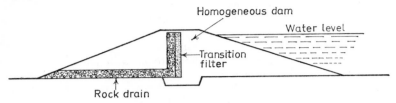

Figure C.4. Chimney drain

chippings Crushed stone fragments ranging from 3 mm to 25 mm ($\frac{1}{8}$ in to 1 in). See AGGREGATE.

chlorination Treatment of sewage and water with chlorine to aid purification; STERILISATION of water with bleaching powder or other substance containing chlorine. Chlorine may impart or generate obnoxious tastes and odours and as a result various modifications of the treatment have been developed. See OZONISATION.

chute A steep conduit or channel for conveying water to a water wheel or to a lower level without causing erosion. See HEAD RACE.

Cipolletti weir A form of MEASURING WEIR with a trapezoidal opening with maximum width at the top and side slopes of 1 horizontal to 4 vertical (Figure C.5). See WEIR.

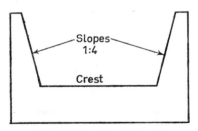

Figure C.5. Cipolletti weir

circle of influence The circular area of ground at the surface corresponding to the circular base of the inverted CONE OF DEPRESSION.

circular-arc method The use of the SLIP SURFACE OF FAILURE principle when investigating the possibility of slides in dam embankments, foundation excavations, and clay banks.

circulating pump Usually a centrifugal pump used to maintain a continuous flow of water through a treatment or cleaning plant or through a condenser of a steam plant.

42

cirque glaciers (or **hanging glaciers**) Small glaciers which are perched well above the valley glaciers and occupy depressions in mountain slopes. The semicircular basins high on the hillsides in the Lake District (U.K.), formerly occupied by small glaciers, are known as CORRIES, CWMS or CIRQUES.

cirques See CIRQUE GLACIERS.

civil engineering Covers all branches of non-military engineering; defined by Thomas Tredgold as 'the art of directing the Great Sources of Power in Nature for the use and convenience of man'. Today, civil engineering deals with roads, bridges, tunnels, dams, railways, land drainage, domestic water supply, irrigation, river control, docks, harbours, water power, sewage disposal, sewerage, airports, and municipal engineering. See FARM WATER SUPPLY ENGINEERING, RIVER ENGINEERING; INSTITUTION OF CIVIL ENGINEERS (Appendix).

clack valve (or **check valve** or **non-return valve** or **reflux valve**) A valve fixed in a pipe or column to allow the flow of water in one direction only. It opens or lifts with the action of the pump to allow water flow and then closes to prevent its backflow; fixed in the pipes leading to or from certain types of pumps (Figure C.6).

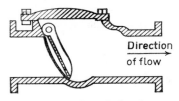

Figure C.6.   Clack valve

clapotis The lapping of waves reflected from a wall, breakwater, or bund, which is above the water level.

clap sill See LOCK SILL.

clarification To clean or remove suspended or colloidal matter from liquids. See CHEMICAL WATER TREATMENT, SUSPENSION.

clay A fine-grained earthy deposit with a grain size less than 0·002 mm and composed mainly of hydrous aluminium silicates with small amounts of other minerals. In water engineering, clay is important as it is practically an impermeable mass and will not permit the passage of water or only at a very slow rate. See SAND, SILT.

clay blanket A construction to reduce the seepage of water through the bed or banks of earthen supply dams or channels. It usually consists of a clay layer, about 0·6 m (2 ft) thick, placed on the slope of the banks or on the bottom of the bed (Figure C.7). See HEELING.

clay core A watertight core or barrier, consisting of puddled clay, in dam construction and embankments. See CUTOFF TRENCH, PUDDLE.

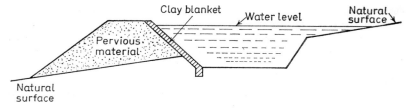

Figure C.7. Clay blanket

clay cutter The shaft-driven bit fitted at the lower end of the suction pipe in a SUCTION-CUTTER DREDGER. It extends the effective range of the dredger considerably.

clay lining See LINING (hydraulics).

claypan A bed of dense, compact, and relatively impervious clay occuring near the surface; stiff or plastic when wet and hard when dry. Its presence may interfere with water movement during the drainage or irrigation of land.

clay puddle (or **puddle** or **puddle clay**) A plastic substance formed by mixing clay with about one fifth of its weight with water. Used as a water-proofing material in hydraulic engineering as under PUDDLE. The coefficient of permeability of puddle clay usually ranges from 3 to 9 mm/yr (0·01 to 0·03 ft/yr).

clay sealing (earthen channels) The injection of clay into earthen channel to seal the voids in permeable sections and thus reduce seepage losses. The treatment has effected seepage reductions of up to 70% and the treatment costs are low. The plant consists of a small centrifugal pump and a drum in which the water is mixed with pulverised dry clay and then spilled down a chute into the channel. See GROUTING.

clay slip The downward movement of a substantial portion of a clayey bank or mountain slope; caused mainly by the presence of water along slippage planes. A clay slip has two manifest characteristics, (1) a prono-unced heave at the base or toe, and (2) tension cracks along the top of the bank. See LANDSLIDE.

cleading A timber lock gate, or the boarding of a cofferdam.

clear-water reservoir See SERVICE RESERVOIR.

climate The average condition of a land area with respect to all atmos-pheric and meteorological influences, mainly of temperature, dryness, pressure, wind, evaporation, light, etc. See ARID, SEMIARID, WEATHER.

climate, continental A climate typical of land areas separated from the moderating influences of the sea by mountain barriers or by distance and marked by relatively large seasonal and daily temperature changes.

climate, oceanic (or **marine climate**) A climate typical of land masses near the oceans, which affect the humidity and modify the temperature and temperature variations.

44

climatic year   A hydrologic and meteorologic record extending over a continuous period of twelve months. See WATER YEAR.

climatology   Science of climate; the systematic measurement and study of climatic elements, including statistical relations, frequencies, mean values, normals, variations, and distribution. See MICROCLIMATE.

clinometer or **batter level**   A hand-held instrument for sighting along inclined planes and for measuring the angle of slope of the sides of earth dams, canals, and embankments. See ABNEY LEVEL.

closed drainage or **non-contributing area**   A land area where surface runoff has no outlet to the sea and flows into lakes, ponds, or other depressions or sinks.

closer   A sheet pile specially cut or fabricated to close a cofferdam in cases when a standard pile will not fit the space.

closing dike   A structure erected across a branch channel of a river to regulate, reduce, or stop the flow of water. A similar structure is sometimes used in conjunction with other works to control the flow in a breakaway.

closure   See STRUCTURE CLOSURE.

cloud   A mass of visible condensed watery vapour suspended in the atmosphere high above general ground level; a cloud at low levels or in contact with the land surface is called MIST or FOG.

cloudburst   A violent rainstorm, in which extremely heavy precipitation falls over a short period. In treeless areas, the runoff cuts deep gullies and often does severe damage to land and property. The streams often become overwhelmed and have their courses changed.

cloud seeding   Under certain conditions, rain can be produced from suitable clouds by seeding them with chemical substances such as silver iodide. A specially equipped aircraft is used.

cloud velocity gauging   A method of measuring velocity in a channel. A known quantity of dye solution is introduced at a station and velocity is measured at the moment a concentration of the solution passes two other given stations located downstream and separated from the injection point by a short turbulent reach. See DILUTION METHODS.

clough   A sluice gate in a CULVERT.

Coal Measures   The Coal Measures of the Carboniferous in the U.K. consist of a great thickness of sandstones, shales, and some coal seams. Some of the sandstones are locally important aquifers, especially for spring water supplies. Although often contaminated, the water is used for industrial purposes and for replenishing canals. Disused coal mines are sometimes important sources of water. See LOWER GREENSAND.

coarse-grained soil   A soil in which sand and gravel predominate and the grains range from 0·5 mm to 2 mm and over. It is a soil type least affected by moisture-content changes, and most surface rain becomes GRAVITATIONAL WATER. See CLAY.

coarse strainers   Coarse drum and bar screens or band strainers placed at water supply intakes to keep out trash and debris and so **prevent**

obstruction or damage to pumps, pipelines, and valves. See MICRO-STRAINERS.

coast The zone between the inland limit of manifest marine influences and the outer limit of breaking waves. See BACKSHORE, FORESHORE, SHORE.

coastal engineering A branch of civil engineering concerned mainly with works to arrest the erosive action of water and waves. See COAST PROTECTION.

coastal lagoon A LAGOON running between the mainland and a sand spit and along which a river flows before gaining an outlet to the sea. Usually it breaks open a new direct outlet only in flood time. In some areas the sand spit is breached periodically to protect the lands around the lagoon from inundation.

coastal plain The area exposed along a coast where the sea retreats from the land mass; the sea floor exposed when the land suffers relative uplift.

coastal plain swamps Uplifted portions of the sea floor which contain an excess of moisture due primarily to their flatness and consequent poor drainage; the water is usually fresh. See DELTA-PLAIN SWAMPS.

coastline The margin of the land along the coast; may comprise a line of cliffs.

coast of emergence The coast where the sea retreats from the land mass as a result of earth movements or EUSTATIC MOVEMENTS.

coast of submergence A coast where the sea advances over the land due to earth movements. The sea enters the valleys and the coast and headlands are more strongly attacked by the waves. See DROWNED VALLEYS.

coast protection The local protection of the coast by arresting the erosive action of water and waves with SEA-WALLS or by GROYNES. The erosion of dunes may be arrested by the planting of shrubs and grasses to bind the surface layer and encourage the growth of a vegetal cover.

coefficient of conductivity See COEFFICIENT OF PERMEABILITY.

coefficient of contraction (C_c) A value related to the VENA CONTRACTA. When a jet of water flows through an orifice under pressure, the area of the jet (A_j) is less than the area of the orifice (A_o). Thus:

$$A_j = C_c A_o$$

where C_c is the coefficient of contraction (Figure C.8).

coefficient of discharge (C_d) The ratio of the actual discharge of water through a weir, pipe, or orifice, to the theoretical discharge. The average value for an orifice is about 0·62. See EFFECTIVE AREA OF AN ORIFICE.

coefficient of imperviousness (U.S.A.) The IMPERMEABILITY FACTOR.

coefficient of permeability The rate of flow of water, usually in litres or gallons per day through unit cross section of soil under a HYDRAULIC GRADIENT of unity and at a specified temperature. Used to calculate seepage in earth channels and dams and also drawdown in wells. Also called COEFFICIENT OF CONDUCTIVITY, COEFFICIENT OF TRANSMISSION, HYDRAULIC CONDUCTIVITY, TRANSMISSION CONSTANT, UNIT OF PERMEABILITY.

coefficient of rugosity See ROUGHNESS COEFFICIENT.

coefficient of transmissibility See TRANSMISSIBILITY COEFFICIENT.

coefficient of transmission See COEFFICIENT OF PERMEABILITY.

coefficient of uniformity See UNIFORMITY COEFFICIENT.

coefficient of velocity The ratio between the actual discharge velocity and the theoretical discharge velocity of water flowing through an orifice. The average value for a sharp-edged orifice is 0·98.

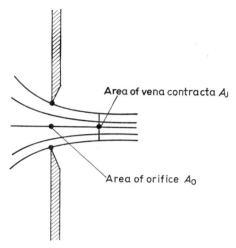

Area of vena contracta A_J

Area of orifice A_O

Figure C.8. Coefficient of contraction

coefficient of viscosity A quantitative expression of the friction between the molecules of a fluid when in motion. The ability of a rock or soil to transmit water varies inversely as the coefficient of viscosity of the water.

cofferdam A temporary watertight case, barrier, or dam forming an enclosure in submerged or waterlogged ground, to exclude water and permit free access to the area within; sheet piling may be used, or a dam built above the ground to exclude water. Cofferdams may be formed to about 9 m (30 ft) below water level; sometimes used in deep wet excavations, foundations, and bridge building. See DOUBLE-WALL COFFERDAM.

cohesive soil A CLAY or clayey SILT, as opposed to SAND which is a cohesionless soil or a frictional soil.

col A depression or pass in a mountain range; a strip of land connecting a range of hills with an outlying feature.

cold front See FRONTAL SURFACE.

cold spring A spring whose water has a temperature below the mean annual temperature of the atmosphere in the area; often used in localities where there are also THERMAL SPRINGS.

collecting area The surface area from which an aquifer receives its supply of water. The area may be large or small and near to, or at some distance from, a well supplied by the aquifer.

collecting system All the drains or sewers in a sewerage system between the buildings from which the sewage originates and the sewage disposal works. See COMBINED SYSTEM.

Collidge dam See MULTIPLE-DOME DAM.

colloidal water treatment The treatment of hard water by the addition of COLLOIDS—often proprietary substances. The colloid forms a film around the solids in the water and prevents their adherence to pipes and metal work. See CHEMICAL WATER TREATMENT, HARDNESS OF WATER.

colloids Substances in a very fine state of subdivision, which are suspended rather than dissolved in water; they diffuse through membrane very slowly or not at all.

colour velocity gauging Determining water velocity by introducing a suitable dye and recording the time it takes to float along a known distance downstream. See CLOUD VELOCITY GAUGING.

columnar sections See VERTICAL SECTIONS.

combination well Applied to an OPEN WELL which is connected to an INFILTRATION TUNNEL or to one or more other wells.

combined system A drainage system in which the surface and soil waters are carried in the same drains and sewers. See SEPARATE SYSTEM.

comb ridges (or **arête**) A rocky spur or sharp-crested ridge between adjacent valley glaciers.

commanded land Generally, land capable of being irrigated by gravity from the outlets of water supply channels.

common excavation A contract term used in earth dam specifications. It includes all material other than ROCK EXCAVATION but includes weathered or soft rock and cemented materials, which are usually ripped for loading purposes. It includes boulders up to 0.76 m^3 (1 yd^3) in volume.

communication pipe A pipe connecting the mains to the boundary of the consumer's property.

compacted yards See METHOD OF MEASUREMENT.

compaction See SUPERFICIAL COMPACTION.

compaction plant Machines fitted with rollers of various types to increase the dry density of earth dams and embankments. The number of passes necessary vary with the soil type, moisture content, and kind of roller used. See PNEUMATIC-TYRED ROLLER, SHEEPSFOOT ROLLER, SMOOTH-WHEEL ROLLER, VIBRATING ROLLER.

compaction, water of See WATER OF COMPACTION.

compensation water The water which must legally be released from a reservoir to meet the requirements of the downstream users who received a water supply before the dam was constructed.

competence (stream) The ability of a stream to carry material; the largest pebble or boulder which a stream can transport. The transporting power of a stream increases at the rate of the 5th or 6th power of its velocity; thus, after heavy rains even boulders are moved along the bed. See TRANSPORT COMPETENCY.

complete flood control See FLOOD CONTROL, COMPLETE.

composite unit graph A graph or tabular presentation covering several unit hydrographs for important sub-areas of a basin or drainage area; allowances are made for time of flow from sub-area outlets to the main gauging station; runoff for each sub-area is computed independently. The graph gives the estimated total flow at the outlet of the basin.

compound coast A coast affected by emergence during one period and by submergence at a later period, both movements being of roughly equal magnitude.

compound cross section A CROSS SECTION of a stream or channel in which the width abruptly increases above a certain level of the sides.

compound dredger A DREDGER equipped with a bucket-ladder for removing mud from the bed of a river or other water and also a CLAY CUTTER. See SUCTION-CUTTER DREDGER.

compound hydrograph A HYDROGRAPH showing flow during an intermittent storm when the effects of one sub-storm continues into the next.

compound weir A flow-measuring structure consisting of a combination of a TRIANGULAR NOTCH and one or more RECTANGULAR WEIRS. Used to determine stream flows over a wide range of dry weather and flood conditions (Figure C.9).

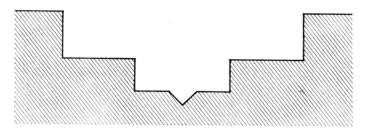

Figure C.9.

compound well A tube well in which the pipes put down into the gravel or sand are of different diameters. See COMBINATION WELL.

concentration The weight of solid matter contained in unit volume of water or other fluid. See SEDIMENT WATER, SUSPENSION.

concrete Construction material consisting of an intimate mixture of cement (which acts as a binder) stone, sand, and water; material hardens to a stone-like mass; used widely for hydraulic structures. See BREEZE CONCRETE, PRESTRESSED CONCRETE, REINFORCED CONCRETE.

concrete bagwork See BAGWORK.

concrete dam All types of dams in which the construction material is mainly concrete, and the term usually includes ARCH DAMS, GRAVITY DAMS, and BUTTRESS DAMS. In view of the cost of concrete, the design is prepared as accurately as the data available allow. Also, the foundations for a concrete dam must possess considerably greater bearing strength than those considered adequate for an earth dam. See PRESTRESSED GRAVITY DAM.

concrete piling Driven concrete piles, used as in TIMBER SHEET piling, as a protection against bank erosion; lasts longer than timber and is not liable to attack by MARINE BORERS. The addition of a stone apron improves the construction but is rather costly. See SHEET PILES.

concrete pipes Plain or reinforced concrete pipes from about 75 mm to 2 m (3 in to 6 ft) in bore and porous when used for SUBSOIL DRAINAGE. They are not commonly used for conveying water under pressure although they can be made to withstand moderate pressures of about 52 m (180 ft). See PRESTRESSED CONCRETE.

condensation (of vapour) Change of vapour into the liquid or solid state; water vapour may change to cloud, rain, dew, sleet, or snow, according to prevailing atmospheric temperature. See DEW POINT.

condensation nucleus The central particle around which condensation of water vapour starts and gathers in the free air.

conduit (water) A general term for any artificial channel, flume, ditch, canal, pipe, or other appliance for conveying water.

cone, alluvial See ALLUVIAL CONE.

cone, debris See DEBRIS CONE

cone of depression (or **cone of exhaustion** or **cone of influence**) The inverted cone-shaped depression in the water-table around a well, shaft, or bore-hole in which pumping is in operation. The rate of pumping and recharge of groundwater will affect the shape and volume of the cone. See CIRCLE OF INFLUENCE, DRAWDOWN, WELL INTERFERENCE.

cone of exhaustion See CONE OF DEPRESSION.

cone of influence See CONE OF DEPRESSION.

confined aquifer See ARTESIAN AQUIFER.

confined eddy See EDDY, CONFINED.

confined water U.S.A. term for ARTESIAN WATER.

confining bed A relatively impervious deposit directly above or sometimes below a water-bearing rock; an impervious rock which confines water under pressure in an underlying aquifer.

confluence The flowing together or uniting of streams, such as the point where a tributary joins the main stream. See BRAIDED COURSE.

connate water (or '**fossil' water**) Native or original water in the interstices of a sedimentary rock which was not expelled during consolidation. See JUVENILE WATER, METEORIC WATER.

consequent streams Streams whose channels were determined by the initial topography or slopes of a newly formed land mass. See DRAINAGE PATTERN, SUBSEQUENT STREAM.

conservation storage Water stored in channels, ponds, and reservoirs during heavy precipitation and later released for power, irrigation, or municipal use.

conservation survey See SOIL CONSERVATION SURVEY, WATER PROSPECT MAP.

consolidation The slow volume reduction of a cohesive soil (clayey soil) under applied loads as in foundations, earth dams, and embankments. During this volume reduction some of the pore water is expelled. See SUPERFICIAL COMPACTION.

consolidation settlement The gradual settlement of a loaded clayey deposit; the process may be accelerated by VERTICAL SAND DRAINS. See SETTLEMENT.

constant rate of injection method A measurement of discharge in an open channel by injecting a constant flow of a solution of known concentration at a station upstream and followed by a dilution measurement at another station downstream where the mixing is complete. See SUDDEN INJECTION METHOD.

construction materials Materials obtained from borrow pits, quarries, and excavations, such as sand, gravel, building stone, road metal, ballast, clay, lime or limestone, etc., and prepared for use in engineering works such as retaining walls, dams, canals, groynes, docks, harbours, and breakwaters. See BORROW PIT, CLAY, MATERIALS SURVEYS, SAND.

consumptive use The amount of water used by living plants plus that transpired and evaporated. See EVAPO-TRANSPIRATION.

contact aerator A tank in which sewage is treated or aerated by injecting compressed air.

contact bed See BACTERIA BED.

contact erosion valley A valley carved by erosion along outcropping lines of weaker rocks at conformable bedding contacts, at igneous contacts, or at upturned unconformities and faults. Valleys formed along fault lines are termed FAULT-LINE VALLEYS, where the block of softer beds becomes the site of the valley.

contact load The material transported by a stream along its bed by sliding and rolling. See COMPETENCE (of stream), LOAD.

contact spring (or **stratum spring**) An outflow of groundwater at or along the base of a permeable bed resting on an impervious bed; often seen at the base of an escarpment. Springs of this type usually give an increased flow after a rainy period. See ARTESIAN SPRING, FAULT SPRING.

continental ice sheets Thick and widespread layers of ice or ICE CAPS in regions where the SNOW LINE is low, such as that covering the Antarctic continent.

continental shelf The submarine continuation of a continent; a submerged platform sloping gently seaward to the CONTINENTAL SLOPE.

continental slope The true boundary of a continental mass with a steep slope downwards to the deep ocean floor.

51

continuity, law of See LAW OF CONTINUITY (flow of water).

continuous delivery (irrigation) A scheme by which each irrigator's allotted quantity of water is delivered at a continuous rate. See ROTATION IRRIGATION.

continuous filter See BACTERIA BED.

continuous stream A stream which flows without interruption or contains water along its entire course between any two points. See INTERRUPTED STREAM.

contour bank A ridge of earth built along a contour line for irrigation or as a soil conservation measure.

contour ditch A ditch or furrow cut along a contour line.

contour furrows Furrows ploughed along the CONTOUR LINES of pasture or other land; they allow water to penetrate the soil but prevent serious erosion, as the water follows the rows at a low velocity. See CONTOUR PLOUGHING.

contour interval The difference in height or elevation between adjacent contour lines. See BENCH MARK, ORDNANCE DATUM, WATER-TABLE CONTOUR PLAN.

contour line (or **contour**) A line drawn on a plan or map to represent an imaginary line on the surface or underground of a definite level at all points. See STRUCTURE CONTOURS.

contour plan A plan showing principally the measured or estimated surface contours of bedrock, water-table, upper surface of an aquifer, or any other deposit of importance during the construction of a reservoir, canal, or other project. At a reservoir site the necessary detailed contouring may be carried out by ground or aerial surveys. The latter are speedy if large-scale Ordnance Survey maps are available and the cost is often lower than with ground surveys. See DAM SITE INVESTIGATIONS.

contour ploughing (or **terracing**) Ploughing furrows at very low slopes along which scour by water is very slight. Adopted as a soil conservation measure on slopes adjoining reservoirs, canals and other works in valleys. See CONTOUR FURROWS.

contracted weir A MEASURING WEIR with a rectangular notch, which is narrower than the channel. If one side of a measuring weir is flush with the channel side it is said to have one END CONTRACTION, or if both sides of the weir are some distance away from the channel sides then it is a weir with two end contractions. See BOTTOM CONTRACTION.

contraction (weir) The combined effect of end contractions and bottom contraction. See CONTRACTED WEIR.

contraction, bottom See BOTTOM CONTRACTION.

contraction, end See END CONTRACTION.

control (1) A stable section in a river or channel, with solid bed and banks, and where the water level is a fairly reliable indication of the flow. (2) A section or area which has received standard treatment or no treatment and is used in research investigations as standard for comparison with an experimental area; a comparative research often used in irrigation investigations. See CONTROL REACH.

control flume Part of an open channel suitable for measuring the flow of water; a short contraction of flume where CRITICAL DEPTH occurs. See RATING FLUME, VENTURI FLUME.

control meter A structure in the channel of a stream where CRITICAL DEPTH is obtained by decreasing the width or raising the bottom, or both combined.

control of flood See FLOOD CONTROL.

control reach A length of channel or stream, upstream from a natural or artificial barrier, along which the water level is relatively stable at some or all stages of discharge. Control may be partial or complete. Partial control exists if downstream fluctuations have some effect on the upstream water level; control is complete if the upstream water level is entirely independent of downstream fluctuations. See CONTROL.

control section See CONTROL REACH.

control valve A valve which regulates the flow of water in a pipeline and gives a constant discharge irrespective of heads or pressures; sometimes applied to a DISCHARGE VALVE.

convectional rain Rain resulting from the upward movement of air heated above the temperature of its surroundings. Thundery rain in a temperate region is commonly typical of convectional rain.

convergence zone, intertropic See INTERTROPIC FRONT.

conveyance loss The loss of water, largely unavoidable, in irrigation and water supply channels. It includes all losses (evaporation, seepage, structure leakage, absorption, etc.) between HEADWORKS and the farms or consumers.

co-operative drainage The joint action of two or more mine owners, landowners, or farmers to drain their properties on the most efficient and economic lines. The cost is shared between the owners on an agreed basis. In mines, pumping stations may be installed at suitable centres, or drain tunnels so positioned as to intersect mine workings below the lowest ore or coal deposit or at some other intermediate level depending on local geology and topography. The cost and length of the drain tunnel and the dewatering benefits are closely investigated. See DRAINAGE DISTRICT, DRAINAGE TUNNEL, LAND ACCRETION.

coppicing A VEGETATIVE PRACTICE for the protection of river and other banks. It involves the cutting back of willows or other vegetation to produce a thick, low growth of branches and suckers which form a dense root growth to bind the soil layers together.

core (1) The cylinder of rock extracted during drilling with the diamond drill or shot-drill. The core is taken from the core barrel and may be about 25 mm (1 in) or more in diameter and up to 6 m (20 ft) or more long. The total length of core recovered compared with the total length of the hole may be from 80 to 95 % depending on the nature of the rocks and the care taken in drilling. The water content of the core is often examined for quality. (2) The above-ground watertight barrier of an earth dam; where it is below ground it is known as the cutoff. See CUTOFF TRENCH.

core trench   A term sometimes applied to the CUTOFF TRENCH.

core wall   See CORE (2).

Cornish pump   A pump introduced early in the 19th century and used widely, especially in mining, for practically 100 years. A single-acting steam engine transmitted the power for pumping through the action of a cumbersome beam.

corrasion   The mechanical work performed by wind and particularly rivers in eroding soils and rocks. The corrasive work of a river is performed mainly by the coarse, angular rock fragments as they are carried and rolled along the bed. See CORROSION.

correction, airline   See AIRLINE CORRECTION.

correction, wet line   See AIRLINE CORRECTION.

corries   See CIRQUE GLACIERS.

corrosion   (1) The chemical wearing away of soils and rocks, particularly by rivers; of minor importance compared with that of corrasion. (2) External corrosion of pipes may be caused by the nature of the ground in which they are placed, or internal corrosion by the aggressive quality of the water carried. The remedies include lining and coating the pipes; using non-metallic pipes; chemical treatment of the water; electrolytic counter-measures, and avoiding corrosive soils. See ENCRUSTATION.

coulee   A deep gully or ravine that flows only in the wet season.

counter-arched revetment   A REVETMENT in brickwork with arches between counterforts similar to a MULTIPLE-ARCH-TYPE DAM. Usually used in a cutting.

counter drain   A drain constructed along the toe of a dam, canal, or embankment to collect and remove seepage and so stabilise the bank.

counterpoise bridge   See BASCULE BRIDGE.

coursed blockwork   Blockwork for breakwater construction in which the 10–50 t precast concrete blocks are laid like masonry in bonded, horizontal courses. See SLICED BLOCKWORK.

course snow   See SNOW COURSE.

coverage (irrigation)   Generally, the maximum depth of water in the farm ditches in relation to the land to be irrigated. A coverage of less than 75 mm (3 in) is not considered adequate for efficient farm irrigation.

covered open channel   See OPEN CHANNEL.

covered reservoir   A SERVICE RESERVOIR enclosed by reinforced concrete roofing or sometimes by prestressed concrete. The roof is often covered with soil.

cover, snow   See SNOW COVER.

cover, soil   The protective mat of vegetation over the soil surface. It retards erosion and runoff and may consist of low-growing shrubs and herbaceous plants.

crack An opening which may develop in an earth dam or embankment due to tensile stresses. See TRANSVERSE CRACK, LONGITUDINAL CRACK.

cradle (1) A structure which supports and is shaped to fit an open or closed conduit. (2) A low steel framework running on rails on a slipway and on which a ship rests and moves during construction, repairs, or launching.

crag Marine and estuarine gravels and sands, referrred to the Pliocene division of geological time, and forming sandbanks or estuarine deposits near the mouth of a large river; well developed in the east of East Anglia (U.K.).

crag and tail A surface feature in which a tail of relatively soft rock is left behind a protecting crag of hard rock. Sometimes seen in glaciated country where a resistant mass of igneous rock lies in the path of the moving ice.

Craig's formula A British flood intensity formula:

$$Q = 440bN \log_e 8L^2/b$$

where Q is the flood intensity in cusecs, L is the greatest length of the catchment in miles, b is the average width of catchment in miles, and $N = Kvi$, where K is the coefficient of runoff, v is the velocity of runoff in ft/s, and i is the average intensity of rainfall in in/h. The coefficient N is a local coefficient to be assessed for the particular catchment (*Proc. Instn Civil Engrs*, **80**, 203).

crater lakes Permanent lakes that occupy depressions of volcanic origin. See CALDERA LAKES.

creek A stream or small river (particularly Australia); a short arm of a stream: a small inlet of sea coast, a bay or cove; any turn or winding.

creep (water) The movement of water and the resulting erosion around and under a structure. See PIPING.

crest (1) The top of a dam, weir, spillway, or bank; often restricted to the overflow portion (Figure B.5). (2) Peak of a flood or wave. (3) The summit of a hill or mountain or the line connecting the highest points along an anticline.

crest gate (or **spillway gate**) A gate located on the spillway crest of a dam to regulate the flood discharge and thus maintain or lower the water level. See FLASHBOARD, ROLLER GATE, SECTOR GATE, SLIDING GATE.

crest level, design See DESIGN CREST LEVEL.

crevasse (1) A gap in the NATURAL LEVEES of a river caused by high water flows (U.S.A.). (2) A crack of considerable depth and width which has developed in a GLACIER.

crib dam (or **crib wall**) A wall, dam, or barrier consisting of precast concrete members or rectangular interlocking timber, forming bays or cells, laid one over the other to a batter of from 1 in 6 to 1 in 8. The cells are filled with stone or other heavy material. See GRAVITY DAM.

crib groynes GROYNES made of logs; may be triangular, rectangular, or any other shape. They are filled with stone and often used in Australia in the

upper reaches of rivers where durable hardwoods and stones are commonly available. See RECTANGULAR CRIB GROYNES, TRIANGULAR CRIB GROYNES.

cribwork Cribwork may be used as revetment along eroding banks or for
small groynes. It consists of boxes made of logs fixed *in situ* and may be
rectangular, triangular, or an other shape. The boxes are filled with stone
or boulders gathered from the river bed. The logs are spiked or wired
together, or sometimes a length of reinforcing steel is passed through bored
holes at the intersection points of the logs. A space, smaller than the size
of the stones, is left between the logs. This causes siltation and also takes
fewer logs. Construction is usually carried out when the river falls to a low
level or where it can be temporarily diverted. The bank may finally be
planted with willow spars. Cribwork has been used in parts of Victoria,
Australia. See HENCOOPING, STONEMESH CONSTRUCTION.

Crimp and Bruges's formula A British formula used to determine velocity
of flow in a sewer:

$$v = 124m^{0.67}\sqrt{i}$$

where v is the velocity of flow, m is the hydraulic mean depth, and i is the
hydraulic gradient. The units are in ft/s and ft. See BARNES'S FORMULA.

critical concentration (irrigation) The limiting amount of an impurity in
irrigation water beyond which it damages the soil and adversely affects
crop growth and also its quality. See SALT BALANCE.

critical depth Applied to a depth of water in a channel corresponding to a
recognised CRITICAL VELOCITY. Sometimes used for the depth of WATER-
TABLE below natural ground surface. See HIGHER CRITICAL VELOCITY.

critical flow A flow condition in a channel whose mean velocity is equal
to a recognised CRITICAL VELOCITY.

critical hydraulic gradient The HYDRAULIC GRADIENT at which a mass of
sand becomes unstable or 'quick' and has a value of approximately unity.
A sand becomes 'quick' if the upward flow of water is sufficient to cause
flotation. See BOIL, QUICKSAND.

critical slope (1) The slope of a channel in which uniform flow occurs at
the CRITICAL DEPTH, and in which the head lost due to friction, etc.,
per unit length is just equal to the fall of the channel in the same length.
Under such conditions the energy gradient will be parallel to the channel
invert. (2) Sometimes applied to the ANGLE OF REPOSE.

critical tractive force (water) The TRACTIVE FORCE of water flow when bed
material commences to move. See BED LOAD, LOAD.

critical velocity There are several 'critical' velocities. The *critical velocity
of Reynolds* is that at which the flow of a liquid ceases to be streamline and
becomes turbulent and where friction becomes approximately proportional to the square of the velocity. KENNEDY'S CRITICAL VELOCITY is that
velocity at which a stream neither deposits nor picks up material when
flowing over an erodible bed. BELANGER'S CRITICAL VELOCITY is that

velocity at which the energy of the flowing liquid reaches its lowest possible value. See HIGHER CRITICAL VELOCITY, STREAMLINE FLOW.

critical velocity, lower See HIGHER CRITICAL VELOCITY.

cross section (stream) A section along a plane at right angles to the mean direction of flow and bounded by the wetted perimeter and the free surface of the stream. See STREAM PROFILE.

cross section, compound See COMPOUND CROSS SECTION.

crown The topmost part of any arch construction, especially of the inside of a sewer or drain. See INVERT.

cruciform groynes Formed by timber pieces about 2 m (6 ft) long fixed together at about right angles to form a series of crosses. The crosses are spaced from 3 to 4·5 m (10 to 15 ft) apart and placed vertically on the river bed and lined parallel to the flow. They are connected with long timber spars wired to the arms of the crosses. The lower ends of the crosses are driven into the river bed to give stability. The groynes are usually protected and supported by stonemesh sausages or drums filled with stone. These are placed on the upstream or downstream side and sometimes on both sides. If available, old rails may be used to construct the crosses. See CRIBWORK, GROYNES.

crustal-movement lakes Lakes that occupy depressions in the land surface as a result of earth movements such as the down-warping of an area. Faulting or subsidence hollows which impound bodies of water would come into this class of lake. See ACCIDENTAL LAKE.

cryology The study of ice in all its forms and effects. See GLACIOLOGICAL SOCIETY (Appendix).

cryopedology The science dealing with intense frost action and ground which is perennially frozen. The term includes the study and application of engineering methods to minimise or avoid the difficulties involved. See PERMAFROST.

cryptoreic Subterranean streams. The term means 'hidden flow' and was coined by Dr Charles Fenner.

culvert Any covered conduit, of box or circular section, for taking a watercourse, drain, or sewer under a railway, roadway, or embankment and usually large enough for a man to pass through. Also applied to a tunnel along which water is pumped to or from a dry dock (Figure S.2). See BOX CULVERT.

cumec A flow of one cubic metre per second. See CUSEC.

cumulative runoff diagram A diagram in which the daily, weekly, or monthly runoff figures are successively added, thus showing the total runoff from the commencement of the record. Cumulative diagrams may be used to obtain the minimum yield available from a reservoir during the period covered.

cumulative volume curve See DISCHARGE MASS CURVE, MASS DIAGRAM.

cup type current meter A CURRENT METER incorporating a cup type element which is rotated on a vertical axis by the water flow. See SPIN TEST.

current   Running water; water flowing rapidly.

current curve   A graph showing the flow of a TIDAL CURRENT. Rectangular co-ordinates are usually used, with velocity shown by the ordinates and time by the abscissae; the flood is taken as positive and the ebb velocity as negative. Broadly, the graph resembles a cosine curve.

current diagram   A graph showing times of slack water and strength, and flood and ebb velocities over a stretch of a tidal waterway. The times are related to tide phases at a specified station.

current difference   The difference between slack water and strength times in a locality and the time of corresponding phase of the current at a station for which predictions are given in a CURRENT TABLE.

current line   A method of measuring velocity of current. A line is graduated and attached to a current pole. The length of line carried out by the current pole in a specified interval of time indicates the velocity expressed in knots and tenths.

current meter   An instrument for measuring water velocity; consists of a wheel with vanes which are rotated by the water when submerged. The velocity is calculated from the counter reading on the meter and the period of submergence. Current meters are usually used in wide rivers and the velocity tests are made at regular intervals and depths across it to obtain the MEAN VELOCITY. The volume of water flowing is equal to the mean velocity multiplied by the cross-sectional area of the water and expressed per minute or other time unit. See INTEGRATION METHOD, STANDARD CURRENT METER.

current meter rating   Current meter tests under controlled conditions to determine the basic relationship between the velocity of water flowing past the meter, and the observed number of electric signals transmitted from the meter per unit of time.

current table   A list of daily predictions of velocities and times of tidal currents; additional predictions may be made for other places by the current differences and constants given.

curtain grouting   Grouted foundation material formed beneath a dam. It consists of a deep, narrow, and continuous zone, and is generally carried out through the base of the CUT-OFF WALL as this helps to make the top of the grout curtain secure. Grouting is done at high pressure through a row of holes usually drilled vertically downwards. See GROUTING.

cusec   One cubic foot per second; usually applied to water flow. 1 cusec = 0·0283 cumecs. See SECOND-FOOT-DAY.

cuspate foreland   A coastal feature which in effect consists of compound spits extending out to sea. Believed to be due to seas approaching the coast obliquely from two main directions and building up successive banks of shingle ridges; they sometimes enclose a small triangular lagoon.

cut   See (1) CUTTING and (2) LOCK CUT.

cut and fill A type of canal construction or other works which is partly below ground in cut and partly above ground in fill or embankment. See BALANCED EARTHWORKS.

cutoff (1) The new channel cut by a river, or excavated, through the neck of an oxbow curve. The natural cut-through often occurs during a flood. An artificial cutoff is often effective in providing a better overall alignment for a river; it diverts the current from bends where heavy bank erosion occurs; it increases velocities and thus promotes local bed scour; it reduces flood levels, and perhaps saves much cleaning and clearing work round a long bend. See PILOT CUTOFF. (2) A construction, such as a wall or collar, of impervious material placed below ground level to reduce water seepage.

cutoff collar (or **anti-seep diaphragm**) A collar, usually of steel or concrete, placed at right angles to a pipeline to prevent water percolating along the outside of the pipe (Figure O.1).

cutoff depth The depth to which CUTOFF TRENCHES are taken below the excavation level.

cutoff ratio Ratio of length of CUTOFF channel to original length of river around loop or meander.

cutoff trench (or **diaphragm wall**) Any trench, wall, or barrier of impervious material placed within and below an earth dam or embankment to prevent or reduce seepage; the construction may be of compacted clay, sheet piling, concrete, or cementation (Figure O.1). See CORE (2).

cutoff wall A relatively watertight wall placed in a hydraulic structure, earth dam, or channel bank to arrest seepage.

cutter-dredger See SUCTION-CUTTER DREDGER.

cutting (or **cut**) Any excavation below ground level in the open for a pipeline, canal, trench, or other works.

cutwater The wedge-shaped head of a bridge pier to streamline the oncoming flow of water.

cut yards See METHOD OF MEASUREMENT.

cwms See CIRQUE GLACIERS.

cycle of erosion The orderly succession of changes in the history of a land surface subject to erosion. A cycle of erosion may be described as young, mature, or old according to the stage reached in its development. Again, according to region and climate, various types of cycles may be recognised, such as the **normal cycle,** where flowing water is the dominant erosive agent, the **arid cycle** in desert regions, and the **glacial cycle** in which ice is the dominant agent.

cycle of fluctuation (or **phreatic cycle**) The total time during which the water-table rises (due to replenishment) and the succeeding fall or decline (due to discharge). It may be daily, yearly or any other period.

cyclic depletion See CYCLIC RECOVERY.

cyclic recovery The gradual rise in level of the main water-table by recharge or replenishment of water in the zone of saturation. The recovery may extend over several years, following a period of cyclic DEPLETION.

cyclone   A violent whirlwind or hurricane of limited extent; strong winds rotating round a DEPRESSION or centre of minimum barometric pressure. See TYPHOON.

cyclonic precipitation   Precipitation associated with areas of low pressure or depressions. See FRONTAL PRECIPITATION, NON-FRONTAL PRECIPITATION.

cyclopean   Concrete aggregate in which the stones are larger than 150 mm (6 in) and sometimes used in the construction of thick concrete dams. See PLUM.

D

Dalmatian type coast   A coast with long narrow inlets and islands running roughly parallel with the coastline, as in Dalmatia on the eastern shores of the Adriatic. See RIA TYPE COAST.

dam   A massive wall or structure erected across a valley or river for impounding water; a bank of earth, mole, or wooden frame to retard the flow or force of water; a natural obstruction or bar, such as ice or landslide debris, across a river. The deposits underlying a proposed dam for impounding water are tested for character and continuity to safeguard against the leakage of water underneath or around the dam. Leakage may be due to faults, joint fissures, solution cavities, buried channels, or porous rocks. Glacial drift is rarely homogeneous and there may be lenses of permeable gravel or sand. If these occur near the structure, seepage of water is likely. The breaking or failure of a dam is sometimes caused by yield or slippage along shale or clay layers in the rock formation on which the structure rests. The entry of water into these clayey layers causes them to soften and slake and they may become slippage planes for the foundation rock. The seepage of water along thin layers of sand, gravel, or clay tends to increase considerably and under the weight of the dam and the pressure of the water the solid rocks in the vicinity may become unstable and the entire structure be endangered. See DAM SITE INVESTIGATION, EARTH DAM, LARGE DAM.

dam, axis of   A reference line marked on plan or section along the length of a dam and used for measuring horizontal dimensions.

dam, crest of   See CREST (1).

dam foundation   Broadly, the lower part of the dam which transmits the weight to the ground and also the valley floor and abutments. River dam foundations are made sufficiently strong to bear the weight of the structure and also water pressure, and made sufficiently tight to prevent seepage under or around the dam. The nature of the valley floor may determine the amount of storage capacity possible or the height of structure that can be safely constructed. Ground conditions are determined by close drilling, trial pits, and other methods. See DAM SITE INVESTIGATION.

dam, groundwater See GROUNDWATER DAM.

damp air Air containing a large amount of moisture or possessing a relative humidity over 85%, as opposed to dry air with a relative humidity below 60%. See DEW POINT, DEW PONDS.

dam site investigation An investigation into the suitability of locations along a valley for the construction of a dam, associated works, and storage area. Some of the important features are (1) a flattish stream gradient which gives a much larger capacity than a steep gradient for any given height of dam (Figure D.1); (2) valley sides which should be close together at the site, so reducing the cost of dam construction; (3) valley sides which

Figure D.1. Dam site investigation

widen out upstream of the dam site, thus making available a large capacity of water storage. Under favourable topographic conditions it is possible to obtain a high STORAGE RATIO. Preliminary schemes may be prepared for various sites and based on the contours given by the Ordnance Survey maps (U.K.) with estimates of yield and cost. See RESERVOIR, STORAGE/EXCAVATION RATIO.

dam, raising of Increasing the height of a gravity or earth dam, either to increase the freeboard in the interests of safety or to increase reservoir capacity. Stressed cables have been used to raise gravity dams, which is cheaper than the usual method of adding concrete to the downstream face. Generally, gravity dams can be raised with greater safety than earth dams.

Darcy's formula A formula for determining the velocity of percolation of water through natural granular materials. It was developed by Darcy in 1856. Thus:

$$V = k(h/l)$$

where V is the velocity of percolation, h is the difference of head under which flow occurs over a distance l, and k is the constant which varies with the nature of the material, and is calculated from field tests.

Field tests may be made to determine the average velocity V. A GROUNDWATER TRACER, such as common salt, is introduced into a well and samples taken from neighbouring wells at timed intervals. Using the sample with the highest salt content, the average velocity of percolation through the ground between the two wells may be calculated. By using this value for V and the hydraulic gradient taken from the water levels in the wells, the value k in Darcy's formula may be calculated.

datum (1) See ORDNANCE DATUM. (2) Any permanent line, plane, surface, or BENCH MARK used or specified as a reference datum for levelling or to which elevations are referred. See MEAN SEA LEVEL.

Deacon diagram A diagram developed by G. F. Deacon about 1902, giving reservoir storage and yield for different annual rainfalls; it was slightly modified in 1949 in C.F. Lapworth's Chart.

deadman An anchor buried in the ground to hold wires or ropes for winches or for holding bank protective works such as stonemesh sausages, logs, etc. May be constructed from heavy timber or concrete and well buried.

dead water zone A stretch, area, or region where water is temporarily restrained.

debacle The sudden breaking up of ice in a river; a rush of water carrying with it broken ice and debris.

debouch To issue from a stream into a plain or open ground; usually from a ravine or wood.

debris cone A fan-shaped heap of sand and debris deposited by a mountain stream when it reaches a plain or valley. See ALLUVIAL FAN.

debris dam A dam or barrier erected across a stream valley for retention of debris, driftwood, gravel, and silt.

decade Ten; a period of ten years. In METEOROLOGY it often signifies a period of ten days.

decrement, groundwater See GROUNDWATER DECREMENT.

deep lead A CONFINED AQUIFER located beneath a PRIOR RIVER.

deep manhole A manhole or INSPECTION CHAMBER with a smaller access shaft above it. See SIDE-ENTRANCE MANHOLE.

deep percolation Rainfall that percolates downwards into storage in subsurface strata. See PERCOLATION, SHALLOW PERCOLATION.

deep well A well more than about 30 m (100 ft) deep, usually with a water-tight casing down to 30 m (100 ft) or so, and thus largely unaffected by surface impurities. It commonly draws water from the zone of saturation and the water is relatively pure, but may be hard. See SHALLOW WELL.

deep well pump In general, applied to a CENTRIFUGAL PUMP located at the bottom of a well and driven by a long shaft from a surface motor. See SINKING PUMP, SUBMERSIBLE PUMP.

defeated stream A stream that fails to maintain its original course due to earth movements and is diverted, ponded back, or even reversed. See ANTECEDENT STREAM.

deficiency The opposite of an excess. Often applied to regions with inadequate rainfall or other hydrological quantities. See ARID.

deforestation The cutting down and clearing away of forests or wooded areas. Haphazard deforestation may be followed by excessive rain-wash and soil erosion and in time the entire mantle of soil may be removed and the ground made useless for cultivation. See AFFORESTATION, SHEET EROSION.

degradation   The lateral and downward erosion by streams; the process in which the land surface is worn down. See ALLUVIATION.

degree of saturation   The volume of voids filled with water, expressed as a percentage of the total volume of voids in a soil or other mass. See POROSITY.

degrees of frost   The number of degrees that temperature falls below freezing point.

delayed junction   Applied to a tributary which is forced to flow along the foot of a natural levee until it can effect a junction with the main stream.

delivery   The volume of water delivered by a pump per unit time, usually expressed in gallons per minute or cubic metres per minute. See DISCHARGE.

delivery box (irrigation)   A structure which controls and splits the available water in the agreed proportions to farm units.

delivery pressure head   The residual pressure of the water at the delivery end of a pipe or at the jet or nozzle in irrigation.

delta   A type of river deposit, roughly semicircular in outline, and formed where a river loses its carrying capacity by flowing into a relatively quiet body of water such as the sea, lake, or reservoir. It consists of three main parts: (1) a broad, gently inclined upper deposit mostly above the stagnant water *(topset beds)*; (2) the submerged and steeper front slope *(foreset beds)*, and (3) the submerged silts and muds which extend outwards as a gently undulating layer from the base of the front slope *(bottomset beds)*. The plan shape of a delta depends on many factors, such as the load carried by the river and the existence of any longshore currents in the body of water entered. Engineering efforts are often made to maintain one of the delta channels in a navigable condition out to sea. The channel selected may be improved by building parallel jetties and inducing the current to scour a deeper channel.

deltaic tract   A PLAIN TRACT of a stream which has extended with the formation of a delta and deposition of fine sediments. See MOUNTAIN TRACT, VALLEY TRACT.

delta-plain swamps   Swamps which owe their existence primarily to their flatness and frequent overflow. In general, the same applies to RIVER FLOODPLAIN SWAMPS. See HILLSIDE SWAMPS.

demonstration area   An area of land, usually a small drainage basin, on which water and soil demonstrational and perhaps experimental work are carried out by and between the local authority and the farmers.

dendritic drainage   A treelike arrangement of rivers and streams, often seen over areas where structural controls are not exaggerated. See DRAINAGE PATTERN.

density   The weight or mass of a substance per unit volume, expressed in kg/m^3 or lb/ft^3. See DRY DENSITY, SPECIFIC GRAVITY, UNIT WEIGHT OF WATER.

density current   The gravity flow of a fluid above, below, or through a fluid of slightly different density, such as the flow of turbid water under clear water in a reservoir or lake.

density, drainage   See DRAINAGE DENSITY.

density of snow   See SNOW DENSITY.

dental   A form of BAFFLE or tooth-shaped projection on a surface or APRON which breaks or deflects the force of flowing water.

dental treatment   The filling of surface cracks, flaws, potholes, faults or weathered rock pockets in the rock foundations of a dam. Concrete and/or grout are used.

dentated sill   A notched sill at the end of an APRON to break the force of a current of water, thereby reducing scour below the apron.

denudation   The wearing away of land, rocks, and soils by mechanical and chemical disintegration, particularly by rain, running water, wind, and the action of frost, snow, and ice.

depletion   The continued discharge or extraction of water from a well, basin, reservoir, pond, or stream at a rate in excess of replenishment. See ECONOMIC YIELD, GROUNDWATER DECREMENT, GROUNDWATER INCREMENT.

depletion (well)   The exhaustion of a water well caused by: (1) interference or excessive pumping by neighbouring wells, (2) depletion of water in storage by pumping in excess of replenishment, and (3) defective casing allowing leakage of water into upper porous rocks. Well depletion may also occur during a prolonged period of drought.

depletion curve (groundwater)   A curve indicating the loss of water from underground storage by springs or seepage into streams.

depletion, cyclic   See CYCLIC RECOVERY.

depletion rate, annual   See ANNUAL DEPLETION RATE.

depletion, stream flow   See STREAM FLOW DEPLETION.

deposit, alluvial   See ALLUVIAL DEPOSIT.

depression   (1) An area of low atmospheric pressure usually surrounded by closed ISOBARS; associated with winds, rain and storm; in Northern Hemisphere winds blow around the depression anti-clockwise and in Southern Hemisphere clockwise. (2) Low-lying land in undulating country within which water collects or flows. See also CONE OF DEPRESSION, DEPRESSION STORAGE

depression, cone of   See CONE OF DEPRESSION.

depression spring   An outflow of water where the surface slopes down and intersects the water-table. See VALLEY SPRING.

depression storage   The volume of water required to fill all natural depressions of an area to their overflow levels; expressed as depth for the entire water surface of the area.

depth area curve   A graph showing relation between progressively decreasing average depth of precipitation over a gradually increasing area, from centre of storm with maximum rainfall outwards to its limits.

depth duration graph A graph which shows the relationship between depth of precipitation and duration of storm.

depth of runoff Total runoff from a basin or drainage area, divided by the area; runoff may be given in depth or volume units per unit of the area.

depth velocity curve (or **vertical velocity curve**) A curve showing relationship between velocity and depth of water at points along a vertical line at a specified cross section of conduit or open channel.

desert Usually, a widespread sandy plain, with an arid climate and largely waterless, treeless, and uncultivated; vegetation is sparse and consists mainly of hardy, drought-resistant types that are able to flourish on the water from a single brief rainy period.

desert peneplain A PENEPLAIN resulting from long-continued desert erosion by wind and water.

desiccation A climatic change with a marked decrease in rainfall, drying up of streams and lakes, and loss of vegetation and surface soil.

design crest level The design level of the top of a dam and ignoring any extra for CAMBER (Figure C.1).

design flood The flood statistics which are used when designing a specific dam, barrage, or other works for the control of a river. The design flood is not necessarily the largest flood possible and the choice is usually made after a study of the hydrology of the river, the damage liable to be caused, and the economic factors involved. See FLOOD (return period).

design flood hydrograph A graph of flow used for studies associated with FLOOD ROUTING and for design of dam or other construction.

design flood level The highest level to which the reservoir surface is estimated to rise in the event of a flood equal to the DESIGN FLOOD.

design recurrence interval The average number of years within which a given hydrological event, such as a flood, is equalled or exceeded, and which serves as a basis for the design of a dam or other hydrologic structure. For example, a return period of 1000 years may be used as a basis for the design of a LARGE DAM. See DESIGN FLOOD, N-YEAR FLOOD.

design storm Measurements and estimations used in determining DESIGN FLOOD; mainly distribution and INTENSITY OF RAINFALL over basin or drainage area.

desilting basin (or **silt basin** or **settling basin**) A basin-shaped enlargement in a stream, where silt and debris carried in suspension may be deposited. See SAND TRAP.

desilting (farm storages) The removal of excess silt and mud from farm storages (ponds and reservoirs) and usually carried out when water levels are low. Some form of mud scoop and two tractors may be used and the material deposited on the downhill side. See DREDGING.

desilting strip (or **filter strip**) A strip of land, located above a reservoir, stock tank, pond, or cultivated fields, and used for the deposition of silt and debris carried by flowing water. The strip contains a vegetal cover of sufficient density to retard flow of runoff water with deposition of suspended material.

detachment (soil erosion)   The disintegration and separation of particles or aggregates from a soil layer usually by flowing water, raindrops or wind; later, this material forms part of the LOAD transported by streams.

detention dam   See DETENTION RESERVOIR.

detention, initial   See INITIAL DETENTION.

detention reservoir   A reservoir for the storage of floodwater or stream flow for a relatively short period; it is retained until the channel or stream can carry released water from reservoir plus its own flow.

detention, surface   See SURFACE DETENTION.

Dethridge meter   A flow meter developed by J. S. Dethridge, M.I.C.E.; used as a basis for payment of water and for measuring flow rate onto the property of an irrigator. The number of revolutions is registered on a counter. The designed maximum flow of a large wheel is 0·113 cumecs (4 cusecs). The small wheel has a designed maximum flow of 0·057 cumecs (2 cusecs). The designed maximum flows given above apply under the conditions of installation in Victoria, Australia.

detrital deposit   A deposit of sand or gravel formed by river action. See ALLUVIAL DEPOSIT, ALLUVIAL FLAT.

detritus chamber (or **detritus tank**)   A tank in which solid matter from sewage settles to the bottom and is removed. See GRIT CHAMBER.

developing a well   Increasing the flow of water from a WELL SCREEN by pumping out the sand and silt to permit easier flow conditions through the coarse material. Gravel may be injected into the ground around the screen. Developing a well may increase the yield of water by up to about 50%.

developmental projects   Schemes intended to improve land conditions, to increase productivity and gain additional revenue from such areas. Would include plans to reduce or prevent flooding of lands bordering rivers, irrigation, and the drainage of swamps, etc. See BENEFIT RATIO.

dew   Liquid water deposited by condensation of water vapour in the air; occurs especially during cloudless, calm nights.

dewatering   The use of a SUBMERSIBLE PUMP or other means to remove water from a shaft or caisson; to remove water from construction excavations or pipe trenches. See DEWATERING HOLE, DRAINAGE, GROUNDWATER LOWERING, PUMP, WELLPOINT.

dewatering hole   A borehole put down to remove or drain away water from existing or proposed sites for cuttings, docks, harbours, tunnels, or dams. The hole may drain water from a near surface deposit downwards through an impervious rock into storage at lower levels. See ARTESIAN WELL (DRAINAGE), DRAINAGE WELL.

dew point   The point at which DAMP AIR, when cooled, commences to deposit its vapour as water or dew.

dew ponds Shallow, often artificial ponds lined with cement or clay; fed by atmospheric CONDENSATION and sometimes rain or drainage of rain; constructed or occur usually on uplands and form a drinking source for cattle and sheep.

diamond-head buttress dam See BUTTRESS DAM.

diaphragm An instrument or appliance for measuring water flow in pipes.

diaphragm pump A RECIPROCATING PUMP in which a rod moves a flexible diaphragm (of canvas, rubber or leather) to and fro. It can deal with very gritty water and is therefore useful in many civil engineering and water projects.

diaphragm wall See CUTOFF TRENCH.

diatomaceous earth. See KIESELGUHR.

diatomite filter A FILTER which extracts fine particles of suspended matter and bacteria by passing the water through elements with a deposited coating of KIESELGUHR or diatomaceous earth. Often used in plants passing small water supplies.

Dickens's formula A British empirical formula expressing the flood intensity as a function of the catchment area; based on the records of a few floods:

$$Q = 825a^{\frac{3}{4}}$$

where Q is the maximum flood intensity in cusecs, and a is the area of catchment in square miles. This formula was devised for Northern India and is considered as approximately correct for districts with an annual rainfall of from 610 to 1270 mm (24 to 50 in). See FLOOD FORMULAS.

differential erosion Applied to the action of flowing water in cutting into the softer rocks at a faster rate than in hard, resistant rocks which remain as ridges or hills. See EROSION.

differential pressure The difference in pressure of a liquid at two different points, as, for example, on the two sides of a meter or device for measuring water flow. See GAUGE PRESSURE, MANOMETER.

differential settlement (dams) When a dam settles in different places by different amounts it is called differential settlement. See SETTLEMENT.

diffuser (1) In pumping, the gradually increasing cross section at the outlet of a centrifugal pump, which reduces the speed of the water and increases its pressure. (2) In sewage, the porous plate or material through which air is blown to stimulate sewage treatment. See CONTACT AERATOR.

diffuser chamber In a TURBINE PUMP, it consists of a number of fixed blades which guide the water outwards, on leaving the impeller, with the minimum of turbulence.

digestion The biological decomposition of organic matter in sewage with the formation of simple and relatively harmless compounds.

dike (or dyke) (1) An embankment, earth ridge, or levee, for confining water especially along river banks to prevent flooding of lowlands. See POLDER. (2) A large drainage channel for land reclamation.

67

dike, closing See CLOSING DIKE.

dilution methods Measurements of water flow in open channels by the injection of solutions of known concentration at points upstream and ascertaining the degree of dilution at points downstream. See CHEMICAL GAUGING, CONSTANT RATE OF INJECTION METHOD, SALT VELOCITY METHOD.

dilutor (irrigation) An appliance fixed to an irrigation water supply pipe for diluting a concentrated stock solution of fertiliser. Usually the dilution rate is adjustable.

dimensionless unit graph (or **non-dimensional unit graph**) A graph plotted in non-dimensional units with respect to flow and time; useful for comparing UNIT HYDROGRAPHS resulting from different storm patterns or of different drainage areas; either constructed from a summation graph or obtained from an observed hydrograph of storm flow.

dip The maximum angle of inclination, or the inclination along a specified direction, of any surface which may be natural or artificial; applied to batters, tunnels, conduits, and other works. See GRADIENT, SLOPE.

Dippel's oil See BONE OIL.

direct-acting pump A RECIPROCATING PUMP in which the power cylinder (compressed-air or steam) and water cylinder are in line and at opposite ends of the same piston rod. See CENTRIFUGAL PUMP.

direct costs (flood protection) The direct expenditure on a flood protection scheme, including all engineering works, loss of land and property, and diversion of transport and other facilities. See FLOOD CONTROL.

direct flood damage All damage caused by flood waters, mainly inundation of land, including stock and property. See FLOOD.

directional drilling The planned drilling of a borehole at an angle off the vertical.

direction float A wooden or metallic float with a small flag to indicate direction of flow of river; the angle that the discharge section line makes with the direction of flow can be measured at an observation station. See FLOAT.

direction of groynes (relative to bank) A GROYNE may be built at right angles to the bank line, slightly upstream or downstream, of that direction. Opinions differ, but in general an IMPERMEABLE GROYNE is erected to point slightly upstream and a PERMEABLE GROYNE pointing a little downstream. A groyne pointing downstream induces siltation below itself, while one pointing upstream tends to cause siltation above and below itself. Groynes inclined slightly downstream are often used in series, so that the area of protection formed by each one extends to the next one downstream.

direct runoff Runoff that has not existed as groundwater or subsurface water since it was precipitated out of the atmosphere. Direct runoff may be expressed as depth of water in mm or inches spread over the entire drainage area or in cubic metres or in acre-feet. See GROUNDWATER RUNOFF.

direct supply A supply line drawing water at mains pressure with no intermediate storage cistern. The method is less costly in pipes and fittings.

disappearing streams Streams that emerge from the hills and flow for a short distance over the valley floor and then sink rapidly into layers of gravel and sand just below the surface and continue as UNDERFLOW. They are fairly common in some arid regions and are found along the Coast Range of California. If the stream is long enough and the seepage and evaporation are abnormal, the entire flow may be used up and the stream then comes to an end. See DRY VALLEY, SINKS, SWALLOW HOLES.

discharge (or rate of flow) The volume of water flowing past a DISCHARGE SECTION LINE in unit time; the delivery rate of a pump or pumping set. Discharge is expressed in litres per second or cumecs or gallons per minute or cusecs. See CURRENT METER, FLOAT MEASUREMENT, GAUGING STATION.

discharge area Area of water flow at DISCHARGE SECTION LINE of stream, channel, or any other waterway.

discharge capacity The safe maximum volume of water which a channel, conduit or any other hydraulic flow structure is capable of passing regularly.

discharge coefficient See COEFFICIENT OF DISCHARGE.

discharge curve A curve showing the relation between DISCHARGE and STAGE of a stream at a given section or station.

discharge head The height between the intake of a pump and the level at which the water is discharged freely to atmosphere. See STATIC DELIVERY HEAD, TOTAL DELIVERY HEAD.

discharge hydrograph A graph in which discharge is plotted as ordinate and time as abscissa. See HYDROGRAPH.

discharge mass curve A curve in which accumulated values of recorded discharges are plotted against intervals of time; the average discharge is represented by the average slope of the curve. See MASS DIAGRAM.

discharge measurement Measuring the volume of water flowing, usually in an open channel. See DISCHARGE.

discharge section line The line between two points fixed on either bank of a stream or channel along which velocities and depths of water are measured. See GAUGING STATION.

discharge site See GAUGING STATION.

discharge valve A valve in a pipeline for increasing or decreasing the discharge of water; frequently placed at the end of the pipeline.

disintegration The breakup or separation of soils and rocks into small fragments by denudation. See DETACHMENT.

dispersion (1) The scattering or separation of soil aggregates into single grains; a factor linked with the erodibility of soils. (2) A SUSPENSION of very fine, often colloidal, particles in a liquid.

displacement (1) The volume of water displaced by a floating vessel or object. See ARCHIMEDES'S PRINCIPLE. (2) The volume of water displaced by a pump ram or piston moving from one end to the other of its stroke.

displacement pump Broadly, any pump which incorporates a ram or piston, but the term is often restricted to an AIR-LIFT PUMP or a DIAPHRAGM PUMP.

dissected peneplain A PENEPLAIN which has been uplifted with respect to the sea level and a new cycle of erosion inaugurated. The streams cut their channels as young valleys below the peneplain which in time is represented perhaps merely by the crests of the hills, which rise to about the same level.

dissection (erosion) The action of water and ice in cutting channels, ravines, and valleys in relatively flat ground.

distributaries Small channels or conduits for conveying irrigation water from the main channels to the farms.

distribution coefficient, rainfall See RAINFALL DISTRIBUTION COEFFICIENT.

distribution curve Showing typical distribution of runoff as percentage of TOTAL RUNOFF from a basin or drainage area.

distribution efficiency See IRRIGATION EFFICIENCY.

distribution reservoir See SERVICE RESERVOIR.

distribution system (1) An irrigation system comprising main channels, irrigation laterals, distributaries, and associated structures for conveying water from the HEADWORKS to the individual farms. (2) Any system which supplies water to a consumer and consists of any or all of the following: service reservoirs, mains, boosters, equipment, and services.

ditch A long narrow channel to hold or conduct water as in drainage or irrigation; a boundary or waterway bordering fields; sometimes called a drain. A field side-ditch has the following minimum dimensions: depth 1 m (3 ft 6 in), top width 1·5 m (5 ft), bottom width 0·5 m (1 ft 6 in). Small dragline excavators, fitted with side-arms, and a specially shaped bucket are sometimes used for ditching work.

ditcher Mechanical equipment for cutting ditches or trenches for pipelines, etc. See TRENCH EXCAVATOR.

ditch, infiltration See INFILTRATION TUNNEL.

ditch outlet Another name for a BAY OUTLET or sluice box.

ditch rider U.S.A. term for WATER-BAILIFF.

diurnal inequality (tidal currents) The difference in velocity between the two flood currents or the two ebb currents of each day; also the difference in height of the two high waters or of the two low waters of each day. See TIDAL RIVER.

diurnal range of tide, great See GREAT DIURNAL RANGE OF TIDE.

diurnal tides Tides with only one high and one low water in each lunar day; tides requiring a full day to complete their cycle. See DURATION OF TIDE.

diurnal tides, semi- See SEMI-DIURNAL TIDES.

diver A person skilled in underwater operations and normally supplied with air by a pipeline from the surface or vessel. Equipped with copper helmet and conventional rubber suit, a diver can work for short periods at a depth of about 70 m (240 ft). Underwater work includes repairs to walls, laying of bridge pier foundations, and salvage of treasure, cargo, and hulls; it may include underwater carpentry, steel plating, and shipwright's work, and the use of the cement gun and cutting and welding tools. See SIEBE DIVING HELMET.

diversion (1) A ditch or channel, with a confining ridge on the lower side, formed across sloping ground to intercept runoff and reduce soil erosion, or to prevent excess runoff from flowing over lands on the lower side. A series of diversions may be constructed in some areas. (2) A channel, cut, or trench to divert water or a stream away or around swamps, dwellings, or engineering works. It may be temporary or permanent. See CUTOFF, PARTIAL DIVERSION.

diversion cut See DIVERSION (2).

diversion dam (or **diversion weir**) A barrier constructed across a stream to divert all or some of the water into another channel or water supply conduit.

diversion requirement See GROSS DUTY OF WATER.

diversion terrace See DIVERSION (1).

diversion tunnel, river See RIVER DIVERSION TUNNEL.

diversion weir See DIVERSION DAM.

diverters or **abstractors** Persons or bodies permitted by the controlling water authority to take water from a stream or water supply channel.

divide (or **drainage divide** or **watershed**) The path along the summit of a range of mountains at which surface water divides and flows down in opposite directions. A major divide would separate river systems rather than individual rivers. See WATERCOURSE.

diving bell A bell-shaped steel chamber, open at the bottom, which can be lowered to or raised from the river or sea bed by a powerful crane. The men within the bell prepare foundations and other works which are necessary underwater. It is now obsolescent as DIVERS are more mobile and also less costly. See HELIUM DIVING BELL.

diving (or **dowsing**) A method used to locate relatively shallow accumulations of water by holding a hazel fork, or other device, in the hands; the free end is claimed to bend downwards when positioned over the body of water in the ground. When submitted to impartial scientific tests, divining has not been successful. See DIVINING DEVICE.

divining device The device used by diviners when locating underground water. It may consist of a hazel, holly, or willow twig, or fencing wire, or watch spring or polished stone suspended by string. Also called dowsing rod, Aaron's rod, Jacob's rod, *baculus divinatorius*, and *virgula divina* (Figure D.2).

division gate A frame or barrier which divides the water flow between two or more LATERALS.

divisor A device which continuously diverts a representative fraction of the runoff from a specified area. The fraction is weighed and sometimes

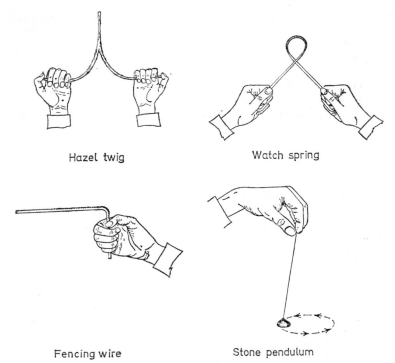

Hazel twig

Watch spring

Fencing wire

Stone pendulum

Figure D.2.   Divining devices

analysed and used to estimate the total soil and water losses for the area (U.S.A.). See DEMONSTRATION AREA.

dock   An artificially formed basin or area of water on the side of a HARBOUR or bank of a river, and closed by gates for the reception of ships. See DRY DOCK, TIDAL DOCK.

docking blocks   Timber blocks used in a DRY DOCK to support the underside of a ship's hull. There are rows of blocks along each side and also a central row of keel blocks. They are made of hardwood, such as oak, and often a softwood cap is fitted to prevent damage to the hull.

dockyards   A naval establishment where ships are equipped, repaired, and maintained.

doldrums   Part of the ocean near the equator with calms and light variable winds, considerable cloudiness, warm moist air, frequent squalls and thunderstorms; a zone of relatively low pressure near the equator. See DEPRESSION.

dolines   Depressions or basins formed by the collapse of large caves in limestone districts. See SWALLOW HOLES.

dolphin A fixed mooring in the open sea formed by a cluster of timber or steel piles driven into the sea bed; a guide for ships entering a harbour with a mouth. See BELL DOLPHIN.

dominant discharge A term with somewhat different meanings according to country or authority. It has been defined as (1) the discharge when a stream is at, or near, its PROFILE OF EQUILIBRIUM; (2) the discharge of a natural channel which determines its depth, width, radius and meander patterns, and (3) the discharge of a natural channel at its BANKFUL STAGE and where the channel is reasonably well defined.

dominant formative discharge A discharge of sufficient volume and in flood season is exceeded a sufficient number of times to affect formation of river bed and course. Sometimes called BED BUILDING DISCHARGE.

dosing chamber See DOSING TANK.

dosing siphon An automatic SIPHON attached to a DOSING TANK for discharging its contents.

dosing tank (or **dosing chamber**) A chamber which receives raw or partly treated sewage; after a certain quantity has collected it is discharged automatically for treatment.

double float See SUBSURFACE FLOAT.

double lock An arrangement of two parallel lock chambers in a canal. They are connected by a sluice which reduces the loss of water to half that with one lock.

double-wall cofferdam A COFFERDAM formed by two parallel lines of sheet piling, used at sites where the height of a single line would be too great for its stability. The space between the lines of piles may be packed with clay for watertightness or with other material if only stability is required.

downstream drains (earth dams) Drains made within earth dams to reduce (a) the pore water pressure in the downstream zone and so increase the stability of its slopes (b) the seepage rate and so prevent PIPING.

downstream total head (over a weir) A measurement at a point downstream of the elevation of the total head relative to the crest. See UPSTREAM TOTAL HEAD.

dowsing See DIVINING.

draft tube The metal casing through which water leaves a turbine. See DIFFUSER.

drain A DITCH (open drain) or a buried pipe (closed drain) for conveying surplus surface or subsoil water or sewage. See CATCHWATER DRAIN, HOUSE DRAIN, SUBSOIL DRAINAGE.

drainage The natural removal of water from land along its streams; the artificial removal of surplus surface water or subsoil water from land by drains; the removal of water by gravity flow from surface works or excavations or irrigation areas; the removal of water or sewage from works or towns. A 'well-drained' soil allows excess water to drain away rapidly, while a 'poorly drained' soil allows excess water to drain only at a slow rate and interferes with plant growth and tillage. See DRAINAGE LEVEL,

73

drainage (dams)   The use of drains or pipes built into the base of a dam to conduct any water up and out through the downstream face. This relieves the upward pressure on the base of the structure due to water in cracks, joints, and bedding planes in the rocks below the dam. The drains are commonly placed near the water face and sometimes inspection tunnels are provided along the length of the dam.

drainage area   See CATCHMENT AREA.

drainage basin   See CATCHMENT AREA.

drainage basin yield   See YIELD OF DRAINAGE BASIN.

drainage, closed   See CLOSED DRAINAGE.

drainage coefficient   Runoff from catchment area or district over a period of 24 hours; expressed in depth or other units.

drainage density   The average length of a river per unit area of its CATCHMENT AREA.

drainage district   (1) A public corporation created under state law to construct, finance, operate, and maintain a drainage system. It is co-operative and self-governing (U.S.A.). (2) Any area which has a DRAINAGE RATE levied on its landholders.

drainage ditch   A channel or ditch cut along the lower part of irrigated land for drainage of excess water. See DRAINAGE.

drainage divide   See DIVIDE.

drainage equilibrium   A hydrogeologic condition in which the volume of water lost to underground storage, from all sources, just equals the volume of water reaching the storage from all sources. A reasonable period of time is taken for the assessment. The position of the water-table is practically unchanged during a period of drainage equilibrium.

drainage, internal   See INTERNAL DRAINAGE.

drainage level   An underground roadway or tunnel with a very slight inclination to divert water and induce its gravity flow outwards and away from mine workings. The roadway is provided with an impervious seal or lining on the lower side to prevent the seepage of water to the dip. See DRAINAGE TRENCH.

drainage outlet   The outlet or lower end of a drainage ditch or other drainage channel.

drainage pattern   The arrangement or layout of the youthful streams draining a valley; it depends mainly on the nature and distribution of the superficial deposits and any surfaces of weakness, such as major joints and bedding planes. See DENDRITIC DRAINAGE, INSEQUENT DRAINAGE, RADIAL DRAINAGE, RECTANGULAR DRAINAGE, TRELLIS DRAINAGE.

drainage rate   A charge or levy upon the owners or occupiers of all lands within a district where a Government, State, or Authority is carrying out drainage works. The rate may be based on the rateable value of the land and may be of such an amount as will provide for interest on the cost of the works, as well as redemption, depreciation, maintenance, and management.

drainage, subsoil See SUBSOIL DRAINAGE.

drainage, surface See SURFACE DRAINAGE.

drainage terrace A graded ridge or embankment of earth used primarily for hillside drainage. It is built with a relatively deep channel and a low ridge. See GRADED TERRACE, NICHOLS TERRACE.

drainage trench A channel or trench cut alongside an underground roadway or tunnel for drainage and to enable the proper ballasting of the rail track. The trench is graded to allow gravity drainage, and may be lined with precast concrete sections on the underside.

drainage tunnel A horizontal or inclined stone drivage in a mine, used mainly for drainage purposes. It may be driven and graded to drain waterlogged rocks, old workings, or near-surface water and thus prevent it seeping downwards to productive workings at lower levels. A central drainage tunnel may serve several mines in close proximity and by gravity flow reduce the need for pumps and pumping stations. A drainage tunnel may be excavated parallel to and a short distance below (to the dip) a main tunnel to take the flow of water, in the Mersey Tunnel connecting Liverpool and Birkenhead (U.K.).

drainage valves Valves provided at all low points of pumping or similar hydraulic systems for allowing water to escape when repairs are carried out or as a frost precaution. The valves are left open when the installation is idle and empty.

drainage well A well put down to drain swampy ground or to remove waste water or sewage at the surface. The water is admitted near the top of the well and discharged into a permeable deposit at lower levels. To be effective, a drainage well must enter a ZONE OF AERATION or a water-bearing rock with subnormal pressure head. See INVERTED CAPACITY, INVERTED WELL.

drain pipes Pipes for the removal of subsoil water, rainwater, waste water, or sewage. They may be made of concrete, earthenware, cast-iron, asbestos cement, or other material and used with open or closed joints. See DRAINAGE, SUBSOIL DRAINAGE.

drain rods Rods pushed to and fro in a drain for cleaning and removing obstructions. They are made in sections which screw together. See LAMP-HOLE, RODDING.

drain tile Applied to AGRICULTURAL DRAIN in U.S.A.

drain well See ABSORBING WELL.

draw A ravine, tributary valley, or coulee, where water discharge is confined to rainstorms; a small natural drainageway; a natural depression; the depth of water required by a ship to float.

drawbridge Any MOVABLE BRIDGE.

draw-door weir A WEIR in which the gates can be raised vertically. See FRAMED DAM, ROLLING UP CURTAIN WEIR, SLIDING-PANEL WEIR, SUSPENDED-FRAME WEIR.

drawdown The lowering or drop in water level of a reservoir; the lowering of the water-table in and around a well or borehole by pumping; usually expressed in metres or feet. See CONE OF DEPRESSION, SUDDEN DRAWDOWN. The term may also be applied when the surface slope of a stream exceeds the normal slope.

drawdown curve The theoretical curvature of the groundwater surface as affected by removal of groundwater by open ditch, tile, or other methods (U.S.A.).

drawdown, sudden See SUDDEN DRAWDOWN.

draw-well A well from which water is extracted by a rope with a bucket attached.

dredger (or **dredge**) A vessel or floating PONTOON equipped with grab, bucket ladder, or suction dredging machine for underwater excavation. In general, harbour dredges are sea-going vessels. A STATIONARY DREDGE is a fixed vessel with a bucket ladder and the excavated material is discharged into a pipeline or a hopper barge. See ELEVATOR DREDGER.

dredging The excavation, bringing up, and removal of material from the bed of a river, lake, or harbour by dredger, dragline, or scoop. Dredging may be necessary to increase the depth of water in a river or to open up an intake channel. Draglines may be used from each bank of a river and the spoil placed directly on the banks. Lakes and other storages may be cleared by scoops pulled across by wire ropes attached to, and operated by, portable power winches.

dredging well The opening in a dredger through which the suction pipe or bucket ladder passes to the river or sea bed.

drift (1) The movement or speed of a current of water or any contained sediment or material. (2) The downstream movement of a discharge boat, anchored or not, during the period taken for velocity measurements. See MODIFIED VELOCITY. (3) See GLACIAL DRIFT.

drift barrier An open structure erected across a stream to catch driftwood; may range from a simple wire fence or chains to massive piers supporting heavy cables between them. The driftwood is removed or burnt at low water. See DEBRIS DAM.

drift-dam lake See GLACIAL LAKE.

drift deposits See BURIED CHANNELS, ENGLACIAL MATERIAL, GLACIAL DRIFT, SHORE DRIFT.

drift, littoral See LITTORAL DRIFT.

drift map A geological map which shows the glacial drift or superficial deposits of an area in addition to solid rocks if they outcrop at the surface.

drift velocity The rate of movement in a specified direction due to drift. See DRIFT (2).

driller's log (or **operational log**) A form containing information entered by the drill foreman for each shift. It covers drilling time, depth drilled, depth of groundwater standing in hole, nature of rocks, colour of sludge, caving of hole, size of casing and whether casing is following the drilling, and any abnormal features or mishaps.

drip irrigation See TRICKLE IRRIGATION.

dripping fault A fault along which small quantities of water seep downwards. Mine workings near such a fault often cause an increased flow due to subsidence and opening of the fault fissure.

driven well In U.S.A. the term is applied to a well driven by a casing at the bottom of which a drive point is fixed. The well is usually between 25 and 75 mm (1 and 3 in) in diameter and there is no drilling, boring, or jetting device.

drizzle Fine, spray-like rainfall; droplets are so small that their individual impact on a water surface is not perceptible.

drop A structure in a channel for lowering the water level, usually vertically but an inclined chute may be used; a steep part of a pipe or channel. See CHECK AND DROP, CHUTE.

drop bars (Australia) See FLASHBOARD.

drop-down curve The curved longitudinal profile of a water surface upstream from a point where a sudden fall occurs in a conduit, river, or channel.

drop-down section The part of the longitudinal profile of a river or channel along which DROP-DOWN CURVE occurs.

drop-inlet dam A dam in which the overflow is carried off through a gently inclined pipe (or SPILLWAY) under the dam. The pipe is connected to a 'riser' or open-topped, vertical pipe at the pond side of the dam.

drop-inlet spillway A spillway consisting of a pipe built under or through the dam or embankment (Figure D.3). See DROP-INLET DAM, OVERSHOT SPILLWAY.

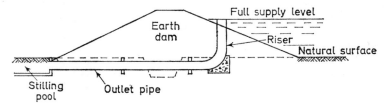

Figure D.3.   Drop-inlet spillway

drop structure A structure built in a stream channel or gully to decrease the velocity of flow by lowering the slope or grade; the drop in velocity promotes the deposition of silt and prevents excessive channel scour. See CHECKDAM.

drought Continuous dry weather, that is, without significant rainfall. See ABSOLUTE DROUGHT; PARTIAL DROUGHT; PRECIPITATION, TRACE OF.

drowned flow The water flow through a flume, or over a weir, is said to be 'drowned' when it is affected by fluctuations in the level downstream. See MODULAR FLOW.

drowned valleys Valley or river mouths which were submerged as a result of land subsidence (or rise of sea-level) and now form long narrow inlets or estuaries. If the valley mouths were broad, the sinking would give rise to wide shallow bays which, as a later phase, may be silted up to form land again. There are many examples of drowned valleys in Britain, as along the south coast of Devon and Cornwall. They are sometimes called RIAS. See COAST OF SUBMERGENCE.

drowning ratio See SUBMERGENCE RATIO.

drum gate A type of spillway CREST GATE in the shape of an acute circular sector in cross section and controlled by valves (Figure D.4).

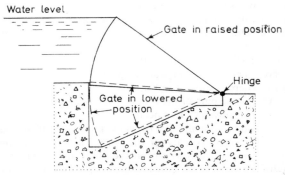

Figure D.4. Drum gate

drum screen A cylindrical or truncated cone-shaped screen, revolving on its centre line, and used in INTAKE WORKS or sewage treatment. See BAND SCREEN.

dry air See DAMP AIR.

dry density The weight or mass of a dry sample of soil per unit volume. Before weighing, the sample is heated at a temperature of 105 °C (221 °F).

dry density/moisture ratio The relationship between DRY DENSITY and moisture content of a soil for a given amount of compaction. The values may be shown as a graph from which the OPTIMUM MOISTURE CONTENT may be determined.

dry dock (or graving dock) An enclosed basin from which the water can be pumped out. A ship is waterborne into the dock, the dock gates are closed, and the water is removed. The ship rests on DOCKING BLOCKS while undergoing repairs or cleaning.

dry farming Farming with no irrigation in arid or semi-arid regions.

dry risers One or more vertical pipes, for fire protection, in a building with outlets at different floor levels. They are normally empty but standard fire hoses can be connected to their lower ends in an emergency. See FIRE HYDRANT.

dry salting The build-up of salt in the ROOT ZONE due to inadequate irrigation.

dry valley A valley without streams; may be caused in various ways. It is a feature of the British Chalk districts where the rainfall percolates downwards with no runoff. The original valleys were carved by streams when at higher levels. Other dry valleys were cut by melt waters from snow and icecaps which were situated on the uplands during the Ice Age. Again, a portion of a valley may become dry by RIVER CAPTURE. In the Mallee country of north-western Victoria, Australia, where the rainfall is low and the ground absorptive, there are few defined streams and all surface runoff is temporary. See UNDERGROUND DRAINAGE.

dry valve shaft A shaft forming part of a storage dam and containing a vertical standpipe for connecting the several draw-off pipes with the supply main. The shaft allows access to almost all the pipework and is generally adopted in relatively large dams. See INTAKE TOWER, VALVE SHAFT.

dry weather flow The flow of water in a stream during a dry period. Dry weather flow varies considerably in different areas, depending on the subsoil conditions and, for example, whether the stream is fed from springs. The term is also applied to the flow in a sewer measured during a 24-hour period of dry weather, which generally corresponds to the consumption of water in 24 hours.

dry well (1) (or **soakaway** or **rummel** (Scotland)) A well into which surface water is drained and allowed to soak into the ground. It may be lined with bricks or stones without mortar and can be empty or filled with large stones. (2) That section of a pumping station where the machinery is housed, as opposed to the WET WELL or SUMP which contains water.

dual water scheme A scheme which utilises both underground and surface water supplies. It gives greater flexibility and the differences in the seasonal fluctuations of the two sources may be exploited. For example, when the water level of an impounding reservoir is low, deep wells in the area often continue to yield substantial supplies.

dug well A well put down by manual labour; only picks and shovels are used. See WELL SINKER.

dumb barge See HOPPER BARGE.

dump truck A heavy-duty, off-highway, rubber-tyred, rear-dump motor lorry. The capacity is about 8 m³ (10 yd³) with a loaded weight of 20–25 t. Used for general earth or rock handling at dam, canal, and cut-and-fill sites. See MOTOR TRUCK, OFF-HIGHWAY TRUCK.

dump wagon A large-capacity wagon, usually tractor-towed, with bottom-, side-, or end-discharge, with tyred wheels or crawler tracks. Used for transport and disposal of earth or rock at engineering excavations. See EXCAVATOR BASE MACHINE.

duplex pump Usually a stream-driven pump comprising two water cylinders and two steam or power cylinders arranged side by side.

duplicate mains Often applied to the laying of SERVICE MAINS, one on each side of a wide concrete highway, to supply consumers on both sides. No communication pipes are required to cross the highway but capital costs tend to be higher.

duration curve A curve showing the frequency with which given values, such as flow and head of water in a channel, are equalled or exceeded during a certain period. For a river used for power generation the graph gives over a certain period (the abscissa) the percentage of total time, or number of times, that the flow was equal to or exceeded the ordinate. The area beneath the duration curve gives the total quantity (or runoff) which flowed in the specified period. The arithmetic mean flow is the average rating.

duration graph, depth See DEPTH DURATION GRAPH.

duration of rainfall In the U.K. the duration of rain falling at a rate of not less than 0·1 mm (0·004 in) per hour and obtained from self-recording rain-gauges.

duration of tide The duration of EBB TIDE, or falling tide, is the interval of time from high water to the following low water; duration of FLOOD TIDE, or rising tide, is the interval of time from low water to the following high water. For a SEMI-DIURNAL TIDE, they cover a period of 12·42 hours, but the presence of non-tidal flow may modify the duration. Duration of ebb tide in a river is usually longer than the duration of flood tide, owing to fresh-water discharge.

Dutch mattress A MATTRESS or layer of timber and reed to prevent scour on a river or sea bed. See REED, REVETMENT.

duty of water See IRRIGATION REQUIREMENT.

dyke See DIKE.

dynamic delivery head See TOTAL DELIVERY HEAD.

dynamic similarity A principle of similarity attributed to Lord Rayleigh which states that if a scale model of a hydraulic structure operates at a speed corresponding truly with that of the full size structure, then the resistances R, densities d, lengths l, and velocities v possess the following relationship:

$$\frac{R_1}{R_2} = \frac{d_1 l_1^2 v_1^2}{d_2 l_2^2 v_2^2}$$

dynamic suction head See TOTAL SUCTION HEAD.

E

early youth A stage of the erosion cycle often characterised by (1) nearly all land surface well above the level of the major streams; (2) flat, wide interstream divide areas; (3) narrow, steep stream gorges, and (4) inefficient, poorly integrated drainage. See LATE YOUTH.

earth and rockfill dam Defined as a fill dam in which the impervious earthfill material exceeds 50%. The free-draining layers outside the earthfill core have an increasing perviousness towards the outside slopes. See EARTH DAM, GRAVITY DAM.

earth auger A hand-boring tool for putting down shallow test holes in soft deposits with no stones. It usually consists of a twist drill mounted on steel piping and rotated by turning a 'T'-piece attached to the upper end. The material adhering to the screw gives an indication of the ground penetrated. Depths down to about 20 ft (6 m) are possible. Earth augers may be used to fix bedrock or the water-table as a preliminary step to putting down shallow wells in areas which are more or less unexplored. The auger may be power-operated and mounted on a lorry or caterpillar tracks, when greater speeds and depths may be reached.

earth dam (or earthen dam) An artificial ridge or barrier for confining water; formed of earth, or rock and earth, with a core of clay, concrete, or other impervious material; may be constructed with an impervious layer, such as steel plate or concrete, etc., on the water face. See HYDRAULIC FILL DAM, ROCKFILL DAM, SUPERFICIAL COMPACTION.

earthflow See LANDSLIDE, MUDFLOW.

earth-moving plant Mechanical equipment such as dozers, scrapers, trench excavators, and loading shovels which are frequently used on earth dams, ditches, drainage, and channel construction. See TRENCH EXCAVATOR.

earthquake A trembling of the earth's crust caused by the mass movement of rocks often along a pre-existing plane of weakness, such as a fault. Small local earthquakes or tremors may also be produced by new heavy loads on the earth's surface such as the initial filling of a large reservoir. Earthquakes often cause severe damage to water mains and the buckling of pipelines. Again, immense damage has been caused by coastal flooding resulting from earthquakes originating below or near the sea.

earthwork Any excavation, cutting, or artificial dumping or banking of SUPERFICIAL DEPOSITS, as during the construction of dams, canals, trenching, and river improvement works. See BALANCED EARTHWORKS.

ebb channel The channel made in an ESTUARY by the EBB CURRENT; usually of 'S'-shape in a long estuary.

ebb current Return of tide water toward the sea; movement of tidal current down tidal stream or from shore.

ebbing well A well in permeable rocks near a coast; the water level rises and falls with the fluctuations of the tides. Excessive pumping may cause salt water to enter the well and contaminate the supply.

ebb tide (or falling tide) The tide during the period between high water and the following low water. See DURATION OF TIDE.

echo sounder An instrument for measuring the depth of water by recording the times for sounds to be echoed back from the sea bed; times are converted into depths and automatically recorded by the instrument. Also used to determine the depth of water in reservoirs. See SOUNDING.

economic storage A storage which gives the cheapest water. The most economical size of impounding reservoir may be estimated by constructing curves of costs and gross yields for dams at different sites or catchments. Usually the cost of yield falls as the size of the reservoir is increased, reaches a minimum, and then increases again. See STORAGE/EXCAVATION RATIO.

economic yield The maximum rate at which water may be extracted from a well or aquifer over a period without depleting the supply or causing a deterioration in quality. Deep wells in valleys have a relatively greater yield than those in upland areas, owing to greater depth in relation to the intake. See PUMPING TEST.

eddy A small whirlpool; to move in a circular direction in flowing water; caused by irregularities or obstructions in bed or banks of stream. See POTHOLES.

eddy, confined Eddy which is not free to move and spread owing to contact with a structure or obstruction in channel.

eddy flow See TURBULENT FLOW.

eddy, free Eddy which is unrestrained and free to spread and move from a structure or obstruction in the channel.

eddy loss The energy lost by eddies, swirls and impact, as opposed to that lost by friction. The energy loss is converted into heat. See IMPACT LOSS.

effective area of an orifice The actual area of an orifice multiplied by its COEFFICIENT OF DISCHARGE.

effective porosity (or **effective drainage porosity**) The ratio of volume of water in a pervious mass (previously saturated with water) which can be drained by gravity forces, to the total volume of the mass. See SPECIFIC RETENTION.

effective precipitable water See PRECIPITABLE WATER, EFFECTIVE.

effective pressure (soils) The pressure between the points of contact of the soil grains; it increases during consolidation of the soil to a maximum value when consolidation is complete. See FLOW NET.

effective rainfall The surface runoff portion of rainfall. See DIRECT RUN-OFF.

effective size Defined by Hazen in his HAZEN'S LAW as the grain size which is larger than 10% by weight of the soil particles, as shown on the GRADING CURVE of the soil, and is described briefly as the D10 size. See UNIFORMITY COEFFICIENT.

effective snow melt That part of SNOW MELT that becomes runoff and reaches a stream. See SNOW DENSITY.

effective storage In certain large storages, the available water is restricted to the quantity that can be drawn off by gravity through the outlet pipes. This is the effective storage. The water below the outlet pipe is known as the NON-EFFECTIVE STORAGE.

effective velocity (groundwater) The volume of groundwater passing through unit cross-sectional area divided by EFFECTIVE POROSITY of the ground or material. Also known as field, true, or actual velocity. See AVERAGE VELOCITY.

efficiency of farm irrigation The percentage of the total irrigation water supplied to a farm which is available to plant roots. See also IRRIGATION EFFICIENCY.

effluent Flowing out; a stream that does not rejoin the main channel after separation; a stream flowing from a lake or underground storage; flow of sewage from a process plant.

effluent seepage Seepage of water from the ground into a channel or stream as opposed to INFLUENT SEEPAGE.

effluent sewage The flow of sewage away from a treatment plant; the treatment may be complete or incomplete.

egg-shaped sewer An ovoid-sectioned sewer, often made of precast concrete segments. They are placed with the small end downwards as this position ensures satisfactory flow when the sewer is nearly empty.

ejector An appliance based on the AIR-LIFT PUMP for raising sewage to a higher level; compressed air is injected into the sewage pipe.

electrical resistance strain gauge A flat coil of fine wire used in investigations into the movements and strains in underground rocks; based on the change in electrical resistance of the thin wire when stretched under the influence of rock strain. In water engineering, it has been used to investigate the strain developed in both longitudinal and circumferential directions in buried water supply pipes.

electrical tape gauge A gauge consisting of a graduated tape with a weight at the end which is lowered to touch the water surface. An electrical device indicates when contact is made with the liquid. See WIRE WEIGHT GAUGE.

electrochemical gauging Measurement of liquid flow in open channels. See CHEMICAL GAUGING, DILUTION METHODS, SALT VELOCITY METHOD.

electro-osmosis A method of lowering groundwater and particularly applicable to silts. It accelerates natural drainage away from surface works and excavations.

elephant's trunk See HYDRAULIC EJECTOR.

elevated ditch A large ditch or canal carried across a depression or canyon by earthfill crossing, similar to an earthfill dam, at normal grade; sometimes used for irrigation.

elevated flume A flume used to convey water across depressions or low-lying areas.

elevation head See POTENTIAL HEAD.

elevator dredger A DREDGER equipped with a bucket ladder in which the buckets dig the underwater material as well as lifting it. They discharge the load into a hold in the vessel or into a chute which leads to a barge.

embankment (or bank) A ridge of earth or rock shaped to contain river water or to support a canal or other construction. Often erected along river banks to prevent overflow. See LEVEE.

83

encrustation (pipes)   The formation of rust nodules or the deposition of calcium carbonate in imperfectly-protected metal pipes, thus reducing their effective area and carrying capacity. See CORROSION (2).

end contraction   A MEASURING WEIR is said to have end contractions when the sharp upstream edges of the notch sides cause the NAPPE to contract in width. See CONTRACTED WEIR.

energy dissipator   Any structure or material used to disperse or destroy the energy of high-velocity waters. See CHECKDAM.

energy gradient (or **energy slope**)   The gradient or slope of the ENERGY LINE.

energy head   The head or elevation of HYDRAULIC GRADIENT line, plus the VELOCITY HEAD of average velocity of water, at any given section; any base may be used as datum.

energy line   A line joining elevations of ENERGY HEADS of an open conduit.

engineer-hydrologist   An engineer versed in the principles of hydrology and their engineering application. See FARM WATER SUPPLY ENGINEERING, HYDROGEOLOGY.

engineering geology   Geological facts and principles that are usefully applicable to mining, municipal, and civil engineering works. It includes rock behaviour when undermined, tunnels, land drainage, irrigation, water supplies, river control, dams, docks, harbours, water power, sewage disposal, and sewerage. When planning and designing these works, the geological aspects are investigated for safe and economic operation. See RIVER ENGINEERING.

engineering practices   All water and soil conservation methods and engineering works which modify land slopes so that runoff may be reduced, controlled, or diverted and thus diminish soil erosion and improve irrigation. It includes the use of DRAINAGE TERRACES, CONTOUR FURROWS, DAMS, DIKES, DIVERSIONS, etc. See also DRAINAGE, IRRIGATION, RIVER TRAINING, STOCKWATER DEVELOPMENT, VEGETATIVE PRACTICES.

englacial material   The disintegrated rock material held and carried forward in the lower part of a glacier. Later, when the ice melts, the debris and boulders are left stranded on alien land. Englacial material of pre-existing glaciers often forms natural dams in valleys, allowing the formation of lakes which may serve as sources of water supply. See GLACIAL LAKE.

engrafted river system   Streams that unite when flowing over the COASTAL PLAIN to reach the new coast.

enrockment   See RIPRAP.

entrance head   The head of water required to cause flow into a closed or open conduit or hydraulic structure. Entrance head is the sum of the ENTRANCE LOSS and VELOCITY HEAD.

entrance lock   The lock providing entrance for ships to a dock in which the water level is different from that outside.

entrance loss (or **entry loss**)   The head lost at the inlet to a conduit or hydraulic structure due to friction and turbulence of water.

84

entry head See ENTRANCE HEAD.

envelope curve (1) A smooth curve covering all peak values of certain quantities, such as rainfall, runoff, etc., plotted againt other values, such as time, area, etc. In this MAXIMUM ENVELOPE CURVE none of the peak values normally extend above the curve. (2) A smooth curve covering all trough values of the quantities as above. In this MINIMUM ENVELOPE CURVE none of the trough points extend below the curve.

ephemeral stream A stream with its channel positioned above the water-table; its water is derived from precipitation. See PERENNIAL STREAM.

epigene Applied to geological changes or processes active at or near the earth's surface. See HYPOGENE.

equalising bed A layer of concrete or ballast along the bottom of a trench in which pipes are laid.

equalising reservoir (or **balancing tank**) An auxiliary reservoir or storage for excess river water and returned for use during a period of shortage.

equation of continuity See LAW OF CONTINUITY.

equilibrium moisture content The moisture content of a soil at a period when the moisture is stationary.

equilibrium, profile of See PROFILE OF EQUILIBRIUM.

equinoctial rains Rains occurring regularly about the time of the equinoxes in equatorial regions.

equinoctial tides The exceptionally high tides generated at the equinoxes (times at which the sun crosses the equator and day and night are equal). When these tides coincide with severe storms and heavy seas, the coarser deposits along the BACKSHORE are often piled up to form a storm-beach or ridge. See NEAP TIDES, SPRING TIDES.

equinox See EQUINOCTIAL TIDES.

equipotential lines Lines or contours of equal water pressure in the ground near or within an earth dam or other water-retaining structure. A model of the structure may be constructed with the water levels to scale. Glass STAND-PIPES, connected to the soil mass, indicate by the water level in them the pressure at the points to which they lead. The equipotential lines can be plotted from the levels (Figure F.4). See FLOW LINES, FLOW NET.

equivalent pipe A pipe which can be used to replace a given system of pipes with equal head loss for a given flow. The calculation may be made using a pipe flow formula or a NOMOGRAM.

erodible Susceptible to erosion; the material eroded. See EROSIVE.

eroding bank A river, stream or gully bank which is undergoing active erosion by water flow; erosion may also occur by water and sediment flowing down the bank during heavy rains.

erosion The gradual wearing away of land forms, rocks, and soils mainly by the action of flowing water and ice. See ACCELERATED EROSION, CREEP, CYCLE OF EROSION, GEOLOGICAL EROSION, SHEET EROSION, SHORE EROSION, SOIL EROSION, SPLASH EROSION, STREAM-BANK EROSION, WAVE EROSION.

erosion class The erosion pattern of a region; a characteristic group or set of erosion conditions over a specified period and area; often forms part of the land information given on a SOIL CONSERVATION SURVEY map.

erosion, stream bank See STREAM BANK EROSION.

erosive The term usually applied to the agent causing erosion such as flowing water; 'erodible' is preferred when referring to the material eroded, such as soil or rock.

escapage Water which has been diverted into an irrigation system and escapes through spillways or otherwise discarded.

escape A channel or wasteway along which the entire flow of a stream may be discharged. See also BY-WASH.

eskers (or eskars) Long mounds of gravel deposited by streams flowing in channels in the lower part of or beneath a glacier and forming ridges at right angles to the ice-front. Also applied to ridges of water-worn material running across plains or valleys; common in Scotland. See KAMES.

estuarine Pertaining to the tidal mouth of a river.

estuarine flat See MUD FLAT.

estuary The passage or mouth of a large river where the tide meets the river current and the water is brackish. See BAR, EBB CHANNEL, FLOOD CHANNEL, TIDAL CURRENT.

eustatic movements Elevations or depressions of sea level on a vast scale.

evaporating basins Lakes in inland basins, as in south-western Victoria, Australia, with no visible outlet to the sea. These lakes may shrink in dry seasons and swell in wet seasons, but evaporation can usually cope with the inflow and they maintain a reasonably constant area. However, on rare occasions the rainfall may greatly exceed the evaporation rate with flooding of marginal lands and properties. In some cases, outlet channels, cuts, or tunnels have been made to minimise flooding hazards. See LAKE.

evaporation The conversion of surface water into vapour or ATMOSPHERIC WATER; the quantity of water converted into vapour under specified conditions. Evaporation varies greatly according to relative humidity, temperature, character of vegetation, wind velocity, and nature of soil. It goes on more rapidly from clayey than from sandy soils and is greater in cleared areas than in areas covered with forests or vegetation. The amount of evaporation may be estimated from intake and outflow at reservoirs, from water levels in tanks, and from pans filled with soil and sunk flush with the ground. See ANTITRANSPIRANTS, MULCHING.

evaporation area (stream) The area along which evaporation of stream water occurs and consists of the stream surface and the wetted ground along each side.

evaporation discharge The discharge of water, derived from the zone of saturation, by evaporation into the atmosphere; may be SOIL DISCHARGE or VEGETAL DISCHARGE.

evaporation loss   That part of rainfall that returns directly to the atmosphere by evaporation and by transpiration from vegetation.

evaporation opportunity   The ratio of the actual rate of evaporation from water or land in contact with the atmosphere, to EVAPORATIVE CAPACITY under existing atmospheric conditions.

evaporation power   An index of the degree to which a surface or area is favourable or unfavourable to evaporation; equivalent to EVAPORATIVE CAPACITY if referring to unit area exposed parallel to the wind and expressed in the same terms.

evaporative capacity (or evaporativity)   The maximum rate of evaporation which can be developed by a given atmospheric environment from a unit area of wet surface exposed parallel to the wind, the temperature of the surface being at all times exactly equal to that of the surrounding atmosphere.

evaporativity   See EVAPORATIVE CAPACITY.

evaporimeter (or atmidometer or atmometer)   An instrument for measuring EVAPORATION.

evapo-transpiration   The loss of moisture from a soil by EVAPORATION and plant TRANSPIRATION. Evapo-transpiration is viewed by some engineers as an unnecessary term.

evapotron   An instrument capable of measuring directly the evaporation over reservoirs. It was developed by the Meteorological Physics Division, Commonwealth Scientific and Industrial Research Organisation, Australia.

excavated tank   A TANK consisting of a storage basin excavated below ground level. See RING TANK.

excavated volume   The volume of earth or rock excavated to build a water storage such as a reservoir, canal, trench, or foundation, or any works below ground level; usually expressed in m^3 or yd^3. See STORAGE/EXCAVATION RATIO.

excavator   A term which includes a large variety of power-operated digging and loading machines; used widely in excavations for dams, canals, trenches, etc. See TRENCH EXCAVATOR.

excavator base machine   A tracted prime mover to which various types of front-end excavating and lifting appliances can be fitted according to requirements. See MOTOR TRUCK.

exceedance interval   The average number of years within which a given hydrologic event, such as a flood, is exceeded. See RECURRENCE INTERVAL, RETURN PERIOD.

excessive precipitation   See PRECIPITATION, EXCESSIVE.

excessive storms   See PRECIPITATION, EXCESSIVE.

excess water   The water that runs off at or near the surface when precipitation exceeds infiltration plus absorption. See RUNOFF.

exit loss   Head lost at the outlet of a conduit or hydraulic structure due to friction and turbulence of water.

expending beach A beach designed mainly for absorbing wave energy.

exposed intake A type of INTAKE WORKS often used in the U.K. to obtain water from a river or sometimes from a lake. The intake is usually sited on a straight stretch of river with no rapid currents, and its level set below the lowest known river level, while the floors of any buildings are formed above the maximum flood level. See SUBMERGED INTAKE.

exsiccation The drying up or draining dry of an area because of some change which causes loss of water or moisture without a reduction in rainfall. The drying up of marshy ground by drainage or seepage to subsurface excavations or cavities are examples. See DESICCATION.

extinct lakes Lakes which have been naturally obliterated in various ways. See OBLITERATION OF LAKES. The former existence of these lakes is indicated by a natural flat or basin filled with lake deposits and perhaps by old beaches with indications of wave action. See LAKE-FLOOR PLAIN.

extratropical cyclone A cyclone exhibiting great variation in intensity, diameter, and wind strength; found in latitudes higher than the tropics.

F

fabric, reinforcing See REINFORCING FABRIC.

fabridam A barrier of plastic sheet placed across a river and anchored to the river bed. When inflated with air or water it can rise to 6 or 9 m (20 or 30 ft); during heavy floods it is wholly collapsible, when it offers very little resistance to flow. Invented by N.M. Imbertson and Loring E. Tabor of Los Angeles, U.S.A.

facing A layer or cover to protect cuttings, sea walls, or dykes. See LINING.

faggot A fascine or a long cylindrical bundle of brushwood. See FASCINES.

faggotting A covering consisting of brushwood FAGGOTS along the submerged areas of river banks where grass cannot be planted. They are weighed down with stones or other means. Thorn faggots are favoured, as they collect silt and also reduce scour. See FASCINES, MATTRESS, REVETMENT.

failure Broadly, there are two forms in water engineering, (1) structural, due to the inability of an hydraulic structure to physically fulfil its initial purpose, and (2) hydrologic, due to a defect in the hydrologic design (such as the failure of a reservoir to reach its designed capacity due to inadequate runoff from the catchment).

fairlead A metal fixture on a ship or quay to guide the ropes and enable ships to berth safely.

fall (1) A sudden drop in bed level of a stream resulting in WATERFALLS. (2) The STREAM GRADIENT expressed as a fall of so many metres per kilometre or feet per mile.

falling apron principle See UNDERWATER APRON.

falling tide See EBB TIDE.

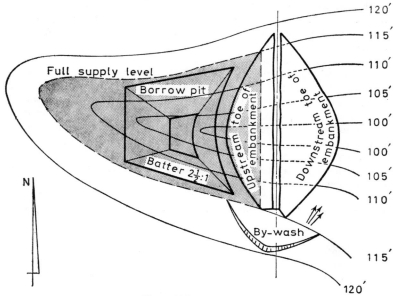

Figure F.1. Farm dam

falls See WATERFALLS.

farm dam A small earth dam constructed on a farm to store water which is released and used as required for irrigation, domestic, or stock supplies (Figure F.1). See BY-WASH, DAM, EARTH DAM.

farm drainage See DRAINAGE, FIELD DRAINAGE.

farm duty See NET DUTY.

farm pond (mainly U.S.A.) A relatively small body of water retained in a hollow excavated in the ground or held behind a dam for farm usage, water for livestock, etc. See GULLY DAM.

farm reservoir A natural or artificial lake near a farm to meet the water requirement of the property such as irrigation and domestic and stock supplies. See STORAGE CAPACITY.

farm water supply engineering A specialised branch of civil engineering concerned with the development and use of the water resources available to farmers and landholders. Much of this service involves the construction of small earthen dams, and the engineer's aid is based on a sound knowledge of soil mechanics and local geology. See WATER HARVESTING.

farm waterway A stream or constructed water channel which is wholly within the boundary of an individual farm.

fascine groyne A groyne, built with shingle and FASCINES, for diverting the current away from an eroding bank. A raft extending out from the

89

bank is secured by cables upstream. Shingle is loaded on to the raft till it sinks evenly. As it sinks, fascines are added along the sides and ends and secured. The box-like structure is loaded with shingle until the raft comes to rest on the bed, and stone or weighed fascines added to close any gaps. See WILLOWBOX GROYNES.

fascines Brushwood which has been rolled into cylindrical bundles and secured with wire; used to protect river banks and sea walls. The brush-wood may be cuttings of ash, hazel, or birch of a suitable age. The material is weighed down or otherwise secured to the bank. Willows or other vegetation are often used in conjunction with fascines. Another application of fascines is in filling ditches or to improve the bearing capacity of soft or waterlogged soils. See ANCHORED TREES, BRUSHWOOD.

fathom A British measure of length equal to 6 ft and used as a unit of marine depth. One fathom = 1·829 m.

fault A break in the rocks along which displacement of one side relative to the other has taken place parallel to the fracture. Faults may form channels for the seepage of water or for the escape of water from a canal or reservoir. The fracture may be sealed off by grouting or by trenching across the fracture and filling the trench with concrete or clay puddle. Faults often form outlets for heavy springs in excavations. Here again, grouting may be applied or a well is sunk and the spring water pumped away clear of the excavation. See EARTHQUAKE, SITE INVESTIGATIONS.

fault coast A coastal feature where the downthrow block of rocks of a fault is covered wholly or in part by the sea and the fault scarp forms the coastline.

fault control (streams) See STREAMS (fault control).

fault-line valley See CONTACT EROSION VALLEY.

fault spring An outflow of groundwater at or along a fault plane. Usually a water-bearing bed is brought against impermeable rocks by the fault and the water flows out at the point where the fault outcrops in a depression or valley (Figure F.2). See FISSURE SPRING.

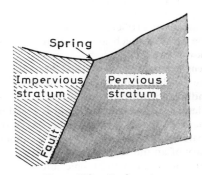

Figure F.2.   Fault spring

90

fault trap (water)   A geological structure in which water in a porous deposit has been trapped or sealed by an impervious deposit thrown opposite it by a fault. See STRUCTURAL TRAP.

feed canal   See CANAL, MAIN IRRIGATION.

feeder   A stream or channel which feeds or supplies water to a canal, reservoir, or pond; a small flow of water from the surrounding rocks into a tunnel, excavation or works. See FAULT, SPRING.

feed-pump   A PUMP which supplies a steam boiler with feed-water; any pump which supplies a machine, appliance, or plant with a regular supply of water of the required purity, quantity, and pressure. See CIRCULATING PUMP.

feed-water   The water pumped into a steam boiler, at boiler pressure, by a FEED-PUMP; the water is often purified, heated to almost boiler temperature, and de-aerated before entry into the boiler; any water which must be regularly supplied to a machine or plant.

Fellenius's circular arc method   A type of landslide of homogeneous earth or clay where failure occurs on a circular arc. See SLIP SURFACE OF FAILURE. Research into this type of failure was undertaken in Sweden after a large portion of Gothenburg harbour had slipped into the sea in 1916. The principle was developed by Petterson and extended by Fellenius in 1927 and is used in the structural design of earth dams.

fender   A shock-absorbing object (ball, rope mat, or old rubber tyre, etc.) hung over the side of a vessel to prevent damage by striking against a pier, quay, or dock or another vessel.

fender pile   Generally an upright, free-standing timber pile driven into the river or sea bed near a berth; it absorbs impact from a berthing vessel.

fender post or **guard post**   See BOLLARD.

fens   See BEDFORD LEVEL.

fetch   The distance which the wind can travel, without obstruction, to any point when raising waves; that is, the distance from the nearest country near the shore in the direction of the wind. See HAWKSLEY'S FORMULA, STEVENSON'S FORMULA.

fidler's gear   Lifting tackle for handling and laying large stones or precast concrete blocks at any angle in the construction of a quay or jetty wall below water level. See BLOCKWORK.

field capacity   The maximum possible amount of water held by a soil against free drainage; the field moisture content of soil two or three days after heavy rain or irrigation. For economic irrigation, the objective is to restore the soil moisture content to field capacity as any quantity beyond this is largely wasted. See MOISTURE REGIME.

field coefficient of permeability   The COEFFICIENT OF PERMEABILITY at temperature of the water.

field drain   See AGRICULTURAL DRAIN.

field drainage (or **farm drainage**)   The removal of water from a field or farm in excess of that required for optimum plant growth and to make the ground firm enough for traffic and cultivation. Field drainage is usually a problem where the soil consists of clays or heavy silts with a low permeability. See DRAINAGE, MOLE DRAINAGE.

field geology   The study and interpretation of rocks in their natural environment and in terms of the economic products sought. It seeks to correlate surface exposures with underground structures. It forms an important branch of HYDROGEOLOGY, particularly in areas which are more or less unmapped.

field moisture   ADHESIVE WATER found above the water-table.

field moisture deficiency   The amount or depth of water required to bring the moisture content of a soil up to FIELD CAPACITY.

field moisture equivalent   The minimum soil moisture content at which a drop of water placed on a smooth surface is not absorbed immediately by the soil but spreads over the surface.

field test   A measurement or experiment carried out under field conditions; control may not be equal to a formal test and the result may be less accurate.

field tile   U.S.A. term for AGRICULTURAL DRAIN.

field velocity   See EFFECTIVE VELOCITY.

fill dam   See EARTH AND ROCK-FILL DAM.

filling   Earth or waste rock deposited to bring construction to the desired formation level, as in embankments, earth dams, canals, rock-fill dams, and similar constructions. See BALANCED EARTHWORKS.

filling, run of the bank   Soil and other material removed indiscriminately from a river bank for use as filling for structures or eroding river banks. The material may be good (e.g. stiff clay) but may also be wasteful and dangerous if it is not secured by protective works or vegetation before the onset of water attacks.

fill yards   See METHOD OF MEASUREMENT.

filter   A stage in a water treatment or purification plant which removes suspended matter. See COARSE STRAINERS, DIATOMITE FILTER, GRAVITY FILTER, PRESSURE FILTER, SAND FILTER, SLOW SAND FILTER.

filter bed   See BACTERIA BED.

filter blocks   Hollow blocks designed to support a BACTERIA BED; made of vitrified clay, often salt glazed.

filter drain (dams)   A drain, about 1·5 m (5 ft) wide, on the downstream side of the clay core of an earth dam to prevent seepage into the fill material.

filter material   A layer of graded granular material which retains solids or suspended particles but allows water to pass through; the broken stone, clinker, or strong metallurgical coke used in a BACTERIA BED. See GRADED FILTER.

filter strip   See DESILTING STRIP.

filter well   A well method of water-table lowering in the vicinity of cuttings or other works to remove water and strengthen the sides. Deep lowering of water may be achieved, since the pump can be submerged in the well. Graded material is provided around the pump suction to prevent fine material being withdrawn with the water. See GROUNDWATER LOWERING.

fines Fine or small particles that settle slowly to the bottom in water. See SLIME, SUSPENDED LOAD.

fiord coast A coast which is deeply indented with long narrow inlets and steep valley slopes which emerge from deep water. In Norway, this type of coast is formed by submerged glaciated valleys.

fire control (catchments) Fire protection of forested catchments; usually includes fulltime firefighting teams, provision of a tower lookout system for early detection, good communications, and also firefighting roads for access. Other firefighting authorities in the area often co-operate in the scheme adopted. See CATCHMENT MANAGEMENT.

fire hydrant A device or arrangement placed in water mains for firefighting purposes. It is controlled by either some form of valve or a wooden ball covered with rubber or composition. A good mains pressure is important. See DRY RISERS.

firn Loose surface snow on heights of mountain ranges. As firn snow becomes older it hardens by a cement of ice. See NÉVÉ.

first bottom A U.S.A. term for the normal FLOODPLAIN of a river. Areas which rarely flood are termed high bottom phase.

fish ladder See FISHWAY.

fish pass See FISHWAY.

fish screen A barrier placed across the inlet or outlet of a pond or channel to prevent the passage of fish.

fishway (or **fish ladder** or **fish pass**) A sloping structure or channel designed to enable fish to ascend or descend a dam, weir, or cataract. See SALMON-LADDER.

fissure spring An outflow of groundwater at or along a fissure, joint, or fault plane. Frequently, several springs occur along the same fissure. The source may be deep, in which case the water is often highly mineralised. See FAULT SPRING.

fissure water Water occurring along cracks, joints, and fissures; a relatively impervious rock may contain a large amount of fissure water. See CHALK.

fixed groundwater Water stored in rocks with fine voids. The water is permanently attached to the pore walls or moves so slowly that it is not available for extraction by wells and pumps. See MICROPORES.

flap trap See ANTI-FLOOD VALVE, FLAP VALVE.

flap valve A valve which allows water to flow in one direction only. See CLACK VALVE.

flashboard (or **stop-log** or **stop-plank** or **drop-bars**) A board, plank, steel joist, or precast concrete beam held horizontally between vertical grooves in end girders, piers, or walls. The beams are placed horizontally on top of each other to close up or control the water level in a spillway or other water channel. See MOVABLE DAM, NEEDLE WEIR.

flashes Lakes occupying depressions in open country: often formed by irregular subsidence of the land by mining operations or removal of underlying salt by natural or artificial means.

flashy stream A YOUTHFUL STREAM which collects from the steep slopes of a catchment area; it has peaks of flood soon after rainfall and the flow subsides rapidly. See NULLAH.

flats See RIVER FLATS.

flat-slab deck dam (or **flat-slab buttress dam** or **Ambursen dam**) A reinforced concrete slab, supported by parallel buttresses, with a flat upstream face inclined at about 45°.

flexible joint A joint which is sealed by rubber rings and has the advantage that the pipe can withstand earth movements and vibration. See RIGID JOINT.

flight of terraces A series of TERRACES or platforms formed by a degrading and swinging stream.

float See FLOAT GAUGE, FLOAT ROD, SUBSURFACE FLOAT, SURFACE FLOAT.

float, captive See CAPTIVE FLOAT.

float, direction See DIRECTION FLOAT.

floaters Isolated masses of rock or ore brought down by a stream from their parent bed and found on the land surface or stream bed. See COMPETENCE (STREAM).

float gauge A small object made of cork, wood, or hollow zinc or copper, which rests and rides on the water surface; its movements are transmitted to a recorder or indicator.

float gauging The measurement of water velocity in an open channel by a float gauge and hence the discharge. See FLOAT MEASUREMENT.

floating crane A large crane, with a lifting capacity up to 200 t, carried on a heavy barge or pontoon. It forms an important appliance in most large ports and docks.

floating dock (or **floating dry dock**) A structure built up of steel plates which may be submerged or floated as required by means of air-chambers. It enables repairs to be carried out on a vessel in the dry. See DRY DOCK, SELF-DOCKING DOCK.

floating dry dock See FLOATING DOCK.

floating foundation See BUOYANT FOUNDATION.

floating harbour An arrangement of PONTOONS connected end to end to form a BREAKWATER.

floating on the system Applied to a reservoir which uses the same pipe to admit and release water.

floating pipeline A pipeline which conveys away the semi-fluid spoil from a SUCTION DREDGER; the pipeline is carried on pontoons and the spoil material may be used in HYDRAULIC FILL DAMS or for land reclamation projects.

floating strainer A bouyant STRAINER or pump suction end which draws relatively clear water from the upper part of the SUMP or water lodge.

A floating strainer is often used when pumping from ponds or rivers where the bed is muddy, but it must be kept below the water level to prevent drawing in air.

float measurement An approximate estimate of the volume of water flowing in a fast stream or open channel by timing a small floating object over a measured distance. A straight, uniform length of channel is selected for the test. The discharge is approximately 80% of the product of the average cross-sectional area of the channel multiplied by the velocity of the float.

float rod (or **velocity rod**) A rod with a weight at the bottom so that it floats in a vertical position; it gives the approximate mean velocity of the water between the bottom of the rod and the water surface; usually timed through a FLOAT RUN.

float rod correction The adjustment sometimes applied to the observed velocity to determine the mean velocity.

float run The known distance between two stations or cross sections, along an open channel or stream, which is used for FLOAT GAUGING.

float, subsurface See SUBSURFACE FLOAT.

float, surface See SURFACE FLOAT.

float switch A switch operated by a float for starting or stopping a pump motor when the water level in the SUMP rises and falls. A similar device may also be used for automatically opening or closing SPILLWAY GATES. See AUTOMATIC PUMP.

float valves Valves installed on inlets to service reservoirs and tanks which close automatically at a predetermined height. See AUTOMATIC SELF-CLOSING VALVES.

flocculation The designed formation of flocs or small masses to accelerate their rate of settlement in water or other liquid; used in sewage or other suspensions. See CHLORINATION.

flood A flow of water in a stream channel which is beyond the capacity of the channel to carry, the excess overflowing the banks to form floodwater; water overflowing land not normally thus covered; any relatively high level or flow of water. Floods may be caused by (1) excessive precipitation; (2) rapid melting of accumulated snow; (3) failure of reservoirs, banks, or levees, and (4) failure of ice jams or bars across rivers. See ANNUAL FLOOD, DESIGN FLOOD, MAXIMUM POSSIBLE FLOOD, MAXIMUM PROBABLE FLOOD.

flood (return period) An approach to the estimation of floods which associates an ever-increasing value of the flood peak with increasing length of the return period. In the U.K. the following classifications are sometimes used: (1) FREQUENT FLOODS—return period of one year; (2) UNUSUAL FLOODS—return period of 10 years; (3) PROBABLE MAXIMUM FLOODS—return period of 50 years; (4) catastrophic floods—return period of 100 years or more. See RETURN PERIOD.

flood absorption The decrease in discharge by accommodation of water flow in a lake, reservoir, channel, or other natural or artificial storage.

flood absorption capacity   The flood storage capacity of a reservoir between high flood level and normal reservoir level.

flood, annual   See ANNUAL FLOOD, FLOOD (return period).

flood axis   The general direction of flow of flood current.

flood channel   The channel formed by the flood tide in an estuary; generally wide at the downstream end and converging to a point at the upstream end. Where the estuary is narrow, the flood channels form a tree-like pattern in plan by branching upstream off the winding EBB CHANNEL.

flood coefficient   An expression, given as a decimal, of the proportion of the PLUVIOGRAPH runoff that actually occurs as surface runoff.

flood control   Concerned with the regulation of flood waters to minimise or prevent the inundation of valuable property or land. Control measures include (1) the building of embankments or levees along the course of the river to confine the flood waters; (2) dredging to deepen the channel and increase its cross section; (3) increasing the gradient of the river by cutting across loops, thus shortening its course, and (4) storing or diverting the flood water. See RIVER TRAINING.

flood control, complete   The adoption of protective measures against large floods, such as MAXIMUM POSSIBLE FLOOD and MAXIMUM PROBABLE FLOOD. See PERMANENT BANK PROTECTION.

flood control storage   A reservoir to reduce the floodwaters downstream of the dam. To impound maximum flood volumes, these special storages are kept as low as possible. The water is later released in controlled discharges which do not usually exceed the capacity of the river.

flood current, tidal   See TIDAL FLOOD CURRENT.

flood damage   The economic loss caused by floodwaters overflowing land and depositing silt and debris. See DIRECT FLOOD DAMAGE, INDIRECT FLOOD DAMAGE.

flood, design   See DESIGN FLOOD.

flood formulas   A rule or statement in algebraic terms for estimating FLOOD INTENSITY. Some are purely empirical and derived from records of actual floods; others are based on theoretical considerations; some again are based on a combination of both of these methods. See CHAMIER FORMULA, CRAIG'S FORMULA, DICKENS'S FORMULA, INGLIS' FORMULA, PETTIS'S FORMULA, RATIONAL RUNOFF FORMULA.

flood frequency   See FLOOD, FLOOD (RETURN PERIOD), FLOOD PROBABILITY.

floodgate   A barrier placed in a water channel to keep out tidal or flood water. See TILTING GATE.

flood hydrograph   A graph showing the FLOOD INTENSITY plotted on a time basis. If the stream is dry before the flood starts then the flood intensity will increase from zero to a maximum in the PERIOD OF CONCENTRATION. A large catchment will give a relatively low flood rate of discharge over a long period and a small catchment a high flood intensity of short duration (Figure F.3).

flood hydrograph, design   See DESIGN FLOOD HYDROGRAPH.

96

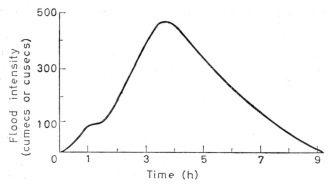

Figure F.3. Flood hydrograph

flood intensity The rate of flood discharge at a given site; usually expressed in cusecs or cumecs.

flood level, design See DESIGN FLOOD LEVEL.

flood losses The water lost to surface runoff due to infiltration, interception, and depression storage during a storm over a catchment area.

flood loss rate The average rate at which water is lost to surface runoff due to infiltration, interception, and depression storage during a storm over a catchment area.

flood peak See PEAK FLOOD.

floodplain The flat lowlands bordering a river and which are partly or wholly covered with water during floods. In some cases, floodplains have a width of many miles or kilometres and extend upstream for great distances. See RIVER TERRACES.

floodplain clays Clays deposited along the lowlands bordering a river during periods of flood. The deposits usually taper out towards the valley slopes and the sediment may be sandy or consist of alternate layers of sand and sandy clay. See ALLUVIAL FLAT.

floodplain discharge The water flowing over the floodplain.

floodplain terraces Natural TERRACES which border many rivers. They may consists of unconsolidated material, or of rock covered with a layer of floodplain deposit, or chiefly of solid rock; when unconsolidated, the material varies from sandy clay to gravel and sometimes boulders. In some areas, the terrace material is sufficiently permeable but retentive to provide water for shallow wells. See INFILTRATION GALLERY.

flood plane The water level of a river during a specific flood.

flood probability The likelihood that a flood of given magnitude will be equalled or exceeded in a given period; probability of 10% would be a 10-year flood, probability of 1% would be a 100-year flood. See FLOOD (RETURN PERIOD).

97

flood protection See FLOOD CONTROL; FLOOD CONTROL, COMPLETE; FLOOD RETENTION BASIN; FLOODWATER RETARDING STRUCTURE.

flood retention basin A dam and reservoir constructed on a river for flood storage. The reservoir is kept empty or partly empty under normal conditions and allowed to fill when a flood occurs. The impounded water is released during and after a flood in such quantities as not to exceed the capacity of the river channel below. Very few rivers contain natural sites favourable for such dams on an economic basis. See FLOODWATER RETARDING STRUCTURE.

flood routing Determining at successive points along a river the timing and form of the HYDROGRAPH of a flood.

flood series A table or list of flood occurrences over a specified period for a river or catchment area. The events are usually arranged in order of intensity or magnitude.

flood stage (1) The level marked at salient points on river banks indicating the elevation above which the river is considered to be in flood. (2) The water level at which a river overflows its natural banks. See BANKFUL STAGE.

flood, standard See STANDARD FLOOD.

flood storage See FLOOD RETENTION BASIN, FLOODWATER RETARDING STRUCTURE, NATURAL STORAGES.

flood strength, tidal See TIDAL FLOOD STRENGTH.

flood tide (or **rising tide**) The tide period between low water and the following high water.

flood tide, duration of See DURATION OF TIDE.

flood warnings Previous information, notice, or signal of the approach of a flood. On large and long rivers, a flood warning can often be given to those along the lower reaches many weeks beforehand, but with swift floods, notice may be very brief. Early flood warnings enable stock and goods to be moved to safety and inhabitants in low-lying areas to be evacuated to higher levels. Warnings are usually based on river heights along the upper reaches, or on rainfall figures at stations in the upper part of the catchment. The information is collected and interpreted by one authority. Automatic river gauges may be connected by land lines to a central authority or to the telephone system. The Australian Commonwealth Bureau of Meteorology has successfully used a model to predict flood flows from isolated storms on several Australian rivers. It is assumed that no significant flood flows occur during a storm until the rainfall has satisfied an initial loss that depends on the MOISTURE STATUS.

floodwater retarding structure A dam built across a small stream or other channel designed to retard and decrease the discharge of floodwater. It has an uncontrolled outlet so that it temporarily impounds runoff from the catchment area and releases it at a rate which prevents the channel overflowing below the dam structure.

floodway   A channel, generally bordered by embankments or levees, designed to carry flood waters. See DIVERSION, LEVEE.

flow   The quantity of water flowing in a channel, stream, conduit, or pipe expressed in volume per unit time. See CUMEC, CUSEC, DISCHARGE.

flowage line (or **flow line**)   A line generally used in connection with the acquisition of rights to flood lands for storage purposes (mainly U.S.A.). It may be a contour line around a pond, lake, reservoir, or the water surface line along a stream and represents a definite level, such as maximum, mean, low, or spillway crest.

flow curve, integrated   See INTEGRATED FLOW CURVE.

flowing well   A well in which the hydrostatic pressure of the water is sufficient to cause it to rise and flow out at the surface. See NON-FLOWING ARTESIAN WELL.

flow lines   (1) The lines shown in a FLOW NET indicating the paths taken by water flowing through an earth dam or the ground near a dam or cofferdam; the lines followed by solid particles in a current of water (Figure F.4). (2) FLOWAGE LINE.

flow meter   An instrument to determine the quantity of water flow in conduits in unit time. See CURRENT METER, INTEGRATING METER, VENTURI METER, WATER METER.

flow net   A net of EQUIPOTENTIAL LINES and FLOW LINES intersecting at right angles in a cross section of a dam. These lines indicate the relationship of head (or potential) and seepage of water through the ground near a dam. They give information regarding (1) ordinary seepage flow, (2) the incre-

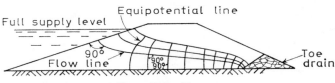

Figure F.4.   Flow net

ased pressure due to seepage after rain, (3) the NEUTRAL PRESSURE at any point from which the upward force or UPLIFT on the dam may be calculated, and (4) the EFFECTIVE PRESSURE on any plane required for dam stability investigations (Figure F.4).

fluctuation of water-table (or **phreatic fluctuation**)   The alternate upward and downward movements of the water-table due to periods of intake and discharge of water in ZONE OF SATURATION.

flume   A wooden, sheet-metal, or reinforced concrete trough or open channel for flushing, sluicing, water power, or conveying water from a higher to a lower level or across depressions or other low-lying areas. See CONTROL FLUME.

flushing   The rapid passage of water through a FIRE HYDRANT or wash-out to prevent the deterioration of water in 'dead-end' mains.

fluvial Pertaining to, or found in, rivers. See FLUVIAL HYDRAULICS.

fluvial hydraulics That branch of engineering science concerned with the motion of water in rivers. See RIVER ENGINEERING.

fluvio-glacial deposits Deposits of detritus formed by GLACIAL STREAMS; material spread out along the margins of ice sheets or at the ice-front. See OUTWASH FAN.

fluviomarine Relating to the action of both rivers and sea, such as the deposits laid down in an estuary or estuarine clays. See MUD FLAT.

fluviomorphology Pertaining to the forms and structures of rivers. See LIMNOLOGY.

fluvioterrestrial The science concerned with land and its rivers.

fly-ash concrete Mass concrete in which pulverised fly-ash (obtained from coal-fired power stations) is used to replace about 20% of the cement content. The temperature rise of fly-ash concrete is about 15 to 20% less than the rise in concrete made with Portland cement. Sometimes used in the construction of dams, particularly hydroelectric dams. See LOW-HEAT CEMENT, TRIEF PROCESS.

flying fox A short form of aerial ropeway, consisting of towers, pulleys, and ropes, for conveying material from one point to another between the towers. Frequently used across a WET GAP or STREAM.

fly-off See INTERCEPTION.

fog Atmospheric obscurity caused by vapour suspended at or near the earth's surface. According to international meteorologic practice, objects one kilometre away are not visible in a fog; when visibility exceeds this distance, but is under two kilometres, the condition is called MIST if caused by water droplets, or HAZE, if by fine dust or smoke. See CONDENSATION.

footing The widening at the base of dams, breakwaters, and similar constructions to distribute the load and improve their stability.

foot irons See STEP IRONS.

foot valve A NON-RETURN VALVE placed at the foot of a column of pipes, such as the suction line of a pump.

forced waves (or **seas**) The waves along the belt of water where they were first generated by wind action. See FREE WAVES.

force pump (or **ram pump**) A pump consisting essentially of a ram or plunger which forces the water to a level considerably higher than itself. On the up-stroke of the plunger, the SUCTION VALVE opens and the water rises in the SUCTION PIPE. On the down-stroke, the suction valve closes and the water is forced through the delivery valve into the discharge pipe.

ford A shallow stretch in a river or other body of water where a crossing may be effected by wading.

forebay A storage reservoir at the head of the penstocks to store water when the load on the hydroelectric plant is reduced and also to supply water for the stages of an increasing load. See PUMPED STORAGE.

foreset beds See DELTA.

forest influences The effects of a large area of trees and plants on water, climate and soil.

foreshore The shore area over which the water's edge migrates during normal tides. See BACKSHORE.

'fossil' water See CONNATE WATER.

foundation The lower part of an hydraulic structure below ground level and which transmits the load to the rock or earth. In a dam foundation, the term includes the valley floor and abutments.

foundation, dam See DAM FOUNDATION.

foundation investigation A branch of SITE INVESTIGATIONS; usually involves the testing by drilling into the deposits below a proposed foundation for a dam or other structure; the estimation of bearing capacities, shear strengths, and settlements and the design of a foundation to suit the prevailing ground conditions form part of the work. See DAM FOUNDATION, DAM SITE INVESTIGATION.

foundation soil The ground or SUPERFICIAL DEPOSITS directly carrying the load of a structure.

fountain An ornamental water jet; a spring or source of river, or a public erection for a regular supply of drinking water, etc.

framed dam (1) Generally a timber dam framed to form a water face and supported by steel struts. (2) A type of movable weir built of steel or cast-iron and timber; some types are lowered to the bed or hung above the water when in flood and propped up by struts against the river bed in low water. See SUSPENDED-FRAME WEIR.

frame weir See FRAMED DAM, SUSPENDED-FRAME WEIR.

Francis turbine A WATER TURBINE in which the water enters radially and leaves axially, with the turbine shaft usually vertical. It operates on a low-head to medium-head and has been installed in a number of HYDRO-ELECTRIC POWER STATIONS.

freeboard The vertical distance between the maximum water level anticipated in flood design and the crest of a dam or other hydraulic structure. Freeboard provides that structures, particularly earth dams, are not over-topped and this provision includes wave action (Figure O.1). See DAMS, RAISING OF.

free eddy See EDDY, FREE.

free flow The flow of water over or through a structure which is not affected by the level of the tail water. See FREE-FLOW WEIR.

free-flowing bore See ARTESIAN WELL (drainage).

free-flow weir (or **free flow** or **free weir**) A weir or dam which is so high that flow over it is in no way affected by the tail-water level. See SUBMERGED WEIR.

free groundwater Groundwater which is not trapped or confined by an overlying impervious rock. See NON-ARTESIAN WELL.

free moisture The moisture in any soil or substance which can be expelled by normal drying.

free water   Water which occupies cavities, fractures, and voids in rocks and is not chemically combined with minerals; water which is free to migrate laterally or vertically and is available for extraction in wells or removed by drainage. See INTERSTITIAL WATER, CONNATE WATER.

free water level   The level or surface of a body of water at atmospheric pressure, that is, in contact or communication with the atmosphere.

free waves (or **swell**)   Waves which have travelled some distance away from the area where they were first formed by wind action. See BREAKERS, FORCED WAVES.

free weir   See FREE-FLOW WEIR.

freeze   A condition when the surface temperature over a widespread area remains below 0 °C or 32 °F for a period long enough to form a characteristic feature of the weather. See SNOW LINE.

freezing rain   Rain containing a portion which freezes to form a film of ice on exposed surfaces.

French drain   An AGRICULTURAL DRAIN in which the pipes are bedded in some filtering material, such as gravel. See GRADED FILTER.

frequency curve   In general, a curve which gives the number of occasions or times that a particular value, event, or phenomenon has occurred during a given period. In hydrology, the curve may show the number of times that quantities like runoff or intensity of rainfall, have occurred or may occur. Usually frequency is plotted along the ordinate and quantity along the abscissa. The graph may indicate frequencies as percentages of the total values or observations available.

frequency diagram   See HISTOGRAM.

frequency of rainfall, intensity   See INTENSITY FREQUENCY OF RAINFALL.

frequent floods   See FLOOD (RETURN PERIOD).

fresh   (1) Water containing less than 1000 dissolved parts of salt per million parts of water (p.p.m.) or milligrammes per litre (mg/l). (U.S.A. Bureau of Reclamation's 1966 Annual Report of Progress of Engineering Research). See BRACKISH. (2) A sudden rush of water in a stream.

freshet   A flood or flow of fresh water into sea; river flood produced by heavy rains or melting snow; particularly a small rush or flow of water of short duration.

fretting   The gradual wearing away of a river bank by water flow; also sometimes by runoff from the adjoining land mass. See EROSION.

friability   The susceptibility of a soil or other mass to crumbling; has a bearing on its resistance to erosion. See ERODIBLE, EROSIVE.

friction   See HYDRAULIC FRICTION.

friction drag   See SURFACE FRICTION.

friction head (or **friction loss**)   The energy or head lost due to HYDRAULIC FRICTION in a pipe or open channel; frequently eddy losses at bends and elsewhere are included.

friction loss   See FRICTION HEAD, HYDRAULIC FRICTION.

friction slope   The FRICTION HEAD or friction loss per unit length of channel or conduit.

fringe water   Water in the ground along the CAPILLARY FRINGE.

frontal apron   See VALLEY TRAIN.

frontal precipitation   The precipitation that occurs at a FRONTAL SURFACE from lifting of warm air over cool air. See NON-FRONTAL PRECIPITATION.

frontal surface   A surface separating two air masses of different characteristics, commonly humidity and temperature, and usually associated with rain and cloud. There are three types: (1) COLD FRONT, along which colder air replaces warmer air, (2) WARM FRONT, along which warm air replaces colder air as the front moves, and (3) OCCLUSION, which is formed when a warm front is overtaken by a cold front, appearing at the surface as either a cold or a warm front.

front-end loader   A power vehicle fitted with a scoop in front for picking up sand, aggregate, gravel, etc., and depositing it above its original position. Sometimes used for construction of groynes and for river works. See EXCAVATOR BASE MACHINE, MOTOR TRUCK.

frost   Freezing weather; frozen dew or vapour; temperature below the freezing point of water; 'degrees of frost' are the degrees that the temperature falls below freezing point. The two main methods of protection of low-growing crops against frost are (1) spraying during the time when frost is actually occurring and (2) applying irrigation not long before a possible night frost. Records of ground frost are obtained from the grass minimum thermometer and earth temperatures by thermometers sunk in the ground at depths varying from 100 mm (4 in) to 3 m (10 ft).

frost action   (1) The alternate freezing and thawing of water held in rock pores, cracks and fissures, resulting in the mechanical breakdown of outcrops. (2) Damage to new construction work, concrete, and mortar when the contained water freezes and expands by about 9% of its volume. Fresh brick or concrete work should be protected from frost action by covering with tarpaulins or other material.

frost-active soil   A soil liable to frost action; undergoes changes in volume and bearing capacity. Some varieties of chalk, silt, and fine sand are commonly frost-active. See PERMAFROST.

frost boil   The loosening or softening of soil caused by thaw after FROST HEAVE.

frost heave   The upward swelling of a soil or pavement by freezing and expansion of the contained water. A silty soil is most liable to frost heave and layers of ice parallel to the ground surface are often formed in the mass. See UPLIFT.

frost line   The line giving the depth of frost penetration into the ground.

frost valve   A small automatic valve sometimes used if there is any danger of a hydrant becoming unusable by the formation of an ice block in the outlet. It ensures that the hydrant outlet or bend is drained and remains empty when not in use. It may be operated by weight or spring.

Froude number A dimensionless number expressing the ratio between influence of inertia and gravity in a fluid. It is the velocity squared divided by length times the acceleration due to gravity. When analysing hydraulic models, the ratio should be similar in both model and full-size plant. See MODEL ANALYSIS.

full meander A stream MEANDER consisting of two loops, one flowing clockwise and the other anti-clockwise.

full supply level (F.S.L.) The water surface level when the reservoir is at maximum operating level, excluding periods of flood discharge. Determined by the spillway crest level where an ungated spillway is provided. Also called TOP WATER LEVEL.

full-tide cofferdam (or **whole-tide cofferdam**) A COFFERDAM in which the height is sufficient to exclude water at all periods of the tide. See HALF-TIDE COFFERDAM.

full-width weir A weir which extends across the full width of the CHANNEL OF APPROACH. See CONTRACTED WEIR.

furrow dams Small earth banks used to retain water in furrows. See BASIN LISTER, DRAINAGE TERRACE.

furrow irrigation The use of furrows or small ditches for conveying irrigation water from a supply ditch or header. When the furrows are placed across the general slope, the method is called GRADIENT IRRIGATION.

G

gabion (1) A wire basket filled with soil or a small CELLULAR COFFERDAM. (2) Rectangular boxes made of wire mesh and filled with heavy stone; used for bank protection in river improvement. If the water level and time allow, a series of steps are cut in the bank and the gabions built up in courses, each course being set back from the one below. Where water is deep, they are made on the bank and rolled down until they reach a footing.

gage See GAUGE.

gallery (1) An underground reservoir or collector for percolating water. (2) A small tunnel constructed in a concrete dam for drainage, access, and inspection. See HEADINGS, TUNNEL.

gallon A measure of capacity, usually for liquids. The British Imperial gallon = $277\frac{1}{4}$ cubic inches = 4 quarts = 4·5459 litres = 10 lb of distilled water. The U.S. gallon = 231·0 cubic inches = 0·833 Imp. gallon. See UNIT WEIGHT OF WATER.

Galloway vacuum tank See PNEUMATIC WATER BARREL.

gang of wells See BATTERY OF WELLS.

gap A short valley across a ridge and connecting lowlands on opposite sides. See WATER GAP.

gas-lift flowing well Introducing natural gas into a well to mix with the water and impart sufficient buoyancy to cause it to rise to the surface.

gas–water surface The surface between an accumulation of natural gas trapped under impervious strata and an underlying body of groundwater.

gate An adjustable frame or barrier to regulate flow in hydraulic constructions such as water channels and dam spillways, and over dam crests. See CREST GATE.

gate chamber (or **camber**) A space or recess provided in a lock wall to receive a SHIP CAISSON or other lock gate when it is open.

gate valve A STOP VALVE with a sliding plate which closes or regulates the water flow in a pipe. When open, the valve offers the minimum of resistance, since flow can occupy the full bore of the pipe.

gathering ground See CATCHMENT AREA.

gauge (or **gage** (U.S.A.)) An instrument for measuring the water level of a river (RIVER GAUGE), or the amount of rainfall (RAIN GAUGE). It may have an automatic recording device. See FLOAT GAUGE, HOOK GAUGE, STAGE.

gauge correlation (or **stage relation**) An empirical curve relating stage or discharge of stream at one or more upstream points to stage or discharge at a downstream point; often used at a downstream point to estimate stage flow at peak.

gauge datum The level of the zero of a gauge relative to a fixed datum or plane of reference. See BENCH MARK.

gauge height See STAGE.

gauge line A line in a fixed direction across a stream or channel and passing through a permanent gauge.

gauge, point See HOOK GAUGE, POINT GAUGE.

gauge pressure The difference between ABSOLUTE PRESSURE and atmospheric pressure; may be positive (pressure gauge) or negative (vacuum gauge). See DIFFERENTIAL PRESSURE.

gauge reading, specific See SPECIFIC GAUGE READING.

gauge, tape See CHAIN GAUGE.

gauge well A chamber sunk in the bank, open at the top, and connected to the stream. It dampens the waves and surges and allows the water level or stage of the stream to be measured in relatively steady conditions.

gauging See CLOUD VELOCITY GAUGING, COLOUR VELOCITY GAUGING, DILUTION METHODS, FLOAT GAUGING, SALT VELOCITY METHOD, STREAM GAUGING.

gauging station A site selected on a stream, river, or open channel and fitted with a GAUGE and other means of making systematic measurements of water level and discharge. The station may be used to record flow conditions for some special purpose, or form one unit in a network to survey the runoff from an entire river basin.

general base level of erosion The level of the sea into which a river flows. See BASE-LEVEL.

geohydrology A term introduced in 1939 by Meinzer who defined it as 'a study of groundwater'; an important requirement in the training of groundwater engineers and geologists. See HYDROGEOLOGY.

geological erosion (or **normal erosion**) The erosion of land forms, rocks, and soils under normal conditions and undisturbed by human activities. See SOIL EROSION.

geological map A map showing the main topographical features of an area, the exposed rocks, and their nature and general distribution. Maps with scales of 6 in. to one mile (1:10 000) are useful for water supply surveys. See TOPOGRAPHICAL MAP, WATER PROSPECT MAP.

geological mapping The location and measurement of rocks, faults, folds, and other geological details exposed at the surface and marking their position on a large-scale topographical map of the area. See BASE MAP.

geology The science which deals with the rocks, and contained minerals and materials, which form the earth's solid crust, and also the agents which produce changes on the surface or underground. The investigation of these materials and processes provides the basic information for geological history.

geomorphology The science which deals with land forms, the distribution of land and water and making geologic interpretations. See HYDROGEO-LOGY.

geophysical prospecting Surface measurements by special equipment to reveal changes in the physical properties of rocks at depth. The information will help to elucidate hidden geological structures and locate, directly or indirectly, some mineral of economic value. The methods found useful are the RESISTIVITY METHOD, the SEISMIC METHOD, and sometimes the MAGNETIC METHOD. Geophysics may be used in the search for water, either by locating rock conditions and structures favourable for the accumulation of water or by utilising the fact that underground water has a direct influence on the electrical conducting properties of the strata. Again, geophysical surveys often make available data relevant to dam and other foundations.

geotechnical engineer An engineer with knowledge of geology and geohydrology and their application to engineering projects and the use of construction raw materials. See below.

geotechnical processes A term suggested in the journal *Géotechnique* in 1946. It covers all processes which change the properties of soils, and also the associated water problems, such as GROUNDWATER LOWERING.

geyser An eruptive spring of hot water and steam; the discharge is usually intermittent. *Geysa* = to gush (Icelandic).

Ghent–Terneuzen Canal A ship canal between Ghent in Belgium and Terneuzen on the Dutch coast, a distance of 35 km (22 miles). It was deepened in 1895 and further enlarged in 1962 to increase its ship capacity from 10 000 to 50 000 tons.

glacial cycle See CYCLE OF EROSION.

glacial drift (or **drift deposits**) Layers of sand, clay, and debris transported and deposited directly or indirectly by glaciers. Since glacial drift may contain permeable beds, lenses, or pockets of sand and gravel, the material is carefully tested at valley sites for dams or similar constructions in order to avoid or minimise leakage. See ENGLACIAL MATERIAL.

glacial drift well A shallow well sunk in GLACIAL DRIFT to obtain water mainly for domestic use and farms, although occasionally the supply is sufficient for municipal purposes. In some localities the structure is favourable for artesian wells. Since the drift deposits are somewhat erratic, the flow conditions tend to vary from point to point. See SHALLOW WELL.

glacial lake (or **drift-dam lake**) A lake formed by a glacier advancing across a river valley and impounding the water. It may also be formed by a TERMINAL MORAINE in a river valley cut prior to the Glacial period. The water is held back by a natural dam of glacial drift deposited across the valley. This type of lake is a hazard on account of floods following the sudden release of the waters. See ORIGINAL CONSEQUENT LAKES.

glacial stream A stream formed by the melting of ice near the snout or termination of a glacier.

glacial valley A 'U'-shaped valley with steep walls formed by vertical erosion by the moving ice mass; the floor shows smooth hollows and irregularities and often contains boulders which are striated and faceted. It is probable that many glacial valleys were originally made by streams and later modified by ice. See CONTACT EROSION VALLEY.

glaciation (1) The wearing down of the earth's surface by moving ice, including the transportation of the material and its later deposition as GLACIAL DRIFT. (2) The process by which large areas of land are covered by ice or glaciers. See ICE AGE.

glacier A thick mass of ice, or snow and ice, formed in regions above the SNOW LINE and moving slowly down valleys or mountain slopes. See HANGING GLACIERS, VALLEY GLACIERS.

glacier burst The sudden downstream rush of water by breakage and removal of glacial ice in a river valley. See ICE GORGE.

glacierets Small glaciers which occupy depressions in mountain slopes, positioned well above the valley glacier. Also called HANGING GLACIERS or CIRQUE GLACIERS.

glaciometer An instrument for measuring the movement of glaciers.

gland A washer or sleeve used in a pump to compress the packing in the STUFFING BOX.

globe valve A water valve in the form of a circular metal disc, which is moved by a threaded spindle. It is open to flow when raised off its seating in the pipe and flow stops when it returns. See GATE VALVE.

go-out A SLUICE formed in a tidal embankment to impound tidal water. At low tide the water can flow out through the sluice.

gorge A narrow channel or passage, with stream, between hills; cut by young streams which erode rapidly downwards and where weathering of the sides proceeds but slowly, as in areas of low rainfall. If the rock forming the sides is hard it will stand with steep or vertical walls. See NARROWS.

gorge, ice See ICE GORGE.

grade (1) The finished surface or bottom of an excavation or canal bed or top of an embankment. (2) The gradient or inclination of a sloping surface or of a line or any horizon; expressed as an angle measured in degrees from the horizontal or as a ratio or a percentage. See STREAM 'AT GRADE'.

graded filter FILTER MATERIAL consisting of coarse gravel, fine gravel, coarse sand, and fine sand, so layered that the fine material will not clog the coarse material. Graded filters are placed along FRENCH DRAINS, at the bottom of excavations which tend to BOIL, and at the downstream toe of an earth dam to improve stability while allowing drainage. Terzaghi gave the following rule to ensure that one layer of the filter will not clog the next. The grading curve of each layer must be plotted. The grain sizes corresponding to 15% of the total of each layer, their D15 sizes (see EFFECTIVE SIZE) are read off. If the D85 size of the finer material is larger than a quarter or at least one fifth of the D15 size of the next coarser layer, there is no risk of the fine material clogging the coarser. See TOE DRAIN.

graded reaches Stretches along a stream where the amount of material supplied is about equal to the amount the stream can transport. The stream is in early maturity and flowing over relatively soft rocks.

graded river floodplain A plain of river denudation with the bedrock concealed by a thin layer of river deposits. The plain is usually narrow and determined by the width of the meander belt. See AGGRADED FLOOD-PLAIN.

graded stream A stream which is neither building up nor actively cutting down its bed. See PROFILE OF EQUILIBRIUM, REGIME.

graded terrace An embankment or ridge of earth having a constant or variable grade or slope along its length. See LEVEL TERRACE.

grader A towed or self-propelled machine with a row of cutting or digging teeth and a blade at the rear to spread and level the material. Used for cutting topsoil and levelling the material. See SCRAPER.

grade-stabilising structure A barrier or dam erected across a stream or gully to stabilise the grade and thus prevent lowering of the channel or further head cutting.

gradient See GRADE, HYDRAULIC GRADIENT, MOISTURE GRADIENT, SLOPE, STREAM GRADIENT, STREAM PROFILE, WATER-TABLE GRADIENT.

gradient irrigation See FURROW IRRIGATION.

grading curve A graph on which the grain size of a sample is plotted on a horizontal, logarithmic scale, and percentages are plotted on a vertical, arithmetic scale. The curve shows, at any selected point, what percentage by weight of particles in the sample is smaller in size than the given point.

grapevine drainage See TRELLIS DRAINAGE.

grassed waterway (or **vegetated channel**) A constructed or natural water channel for carrying surface water from cropland; generally shallow and broad and covered with erosion-resistant grasses.

gravel An unconsolidated deposit consisting of roughly rounded rock fragments with a lower size limit of 2 mm. Conglomerate is the consolidated equivalent. Gravels may or may not be related to the present system of river valleys. The river gravels shown on a map usually occur as TERRACES, each of which is often at a constant height above the present river. FLOODPLAIN gravels, on the other hand, are level with the existing floodplain of the river.

gravel bank See SHINGLE BANK.

gravel filter well A large-diameter well in which graded pea-gravel is used as filter material between two slotted tubes in the well. The outer tube is often withdrawn and the gravel allowed to spread into the ground around the well. To prevent choking of the filters by excessive pumping, a number of wells may be spread over the area to obtain the water volume required.

gravel fraction That fraction of a granular mixture composed of particles between 2·0 and 60 mm in size.

gravel packing (boreholes) The addition of gravel around the outside of a screen or slotted pipe to a thickness of about 75 mm (3 in). Often used in boreholes formed in unconsolidated ground and has the effect of increasing the diameter of the hole and reducing the entry of sand and also the velocity of the water entering the pipe. See GRAVEL FILTER WELL.

gravel pump A type of CENTRIFUGAL PUMP suitable for raising a mixture of gravel and water. Rubber may be used as lining to the pump and pipes which lasts longer than iron or steel. It also has renewable impellers and a renewable lining. See BAILER.

graving dock See DRY DOCK.

gravitational water Water in soils and rocks above the water-table; water which seeps downwards under the influence of gravity. See VADOSE WATER.

gravity-arch dam A dam which maintains its stability by the combined effects of its own weight and arch action.

gravity dam A type of dam in which resistance to overturning, pivoting, or sliding forces is provided by the weight of the dam alone. Such dams are usually of concrete or masonry, of triangular section, and very heavy and costly (Figure A.1). See PRESTRESSED GRAVITY DAM.

gravity filter A sand filter consisting of an open tank in which the difference in level between the filtered water channel and the water in the tank is sufficient to force the water through the sand bed. Since individual units can be built to deal with flows up to 13 500 m^3/d (3 m.g.d.), it is often used in major plants. See FILTER.

gravity groundwater The water which would drain out of a rock in the ZONE OF SATURATION, assuming the zone and the CAPILLARY FRINGE moved downwards for a period, no water entered the area, and none was lost except through the force of gravity. The water discharged as springs and that withdrawn from wells is gravity groundwater.

gravity main Any pipeline from a reservoir which conducts water to lower levels by gravity; usually from the IMPOUNDING RESERVOIR downhill to the SERVICE RESERVOIR. The friction loss of maximum flow through the pipeline must not exceed the difference in level between the top water level of the service reservoir and the bottom water level of the impounding reservoir.

gravity scheme Usually applied to a water supply layout in which the bulk of the water flowing out or into an IMPOUNDING RESERVOIR is by gravity and pumps are not required. See GRAVITY MAIN.

gravity spring Water discharged at the surface from permeable beds under the action of gravity; a spring at the outcrop of the water-table, usually in valleys. Also called DEPRESSION SPRING.

grazing management (catchments) On rangeland, grazing management is concerned with giving a continuing output of animal products while providing that the value of the area as a water catchment is not impaired. Where soil erosion has occurred or is likely to occur, grazing may be deferred for a period or the grazing pressure reduced. See CATCHMENT MANAGEMENT, RANGE.

great diurnal range of tide The difference in height between two low waters and mean higher high waters in each lunar day. See SEMI-DIURNAL TIDE.

Great Ice Age See ICE AGE.

Great Oolite (water supply) This formation in the U.K. consists of sandstones, limestones, marls, and clays. For water supplies, the Great Oolite Limestone is the most important. Many towns and villages obtain their supplies from wells and springs on or near the outcrop. The water varies considerably in quality and quantity, and it may be hard or very hard. See CHALK.

gribble A crustacean found in warm waters; attacks timber by eating into it. See MARINE BORERS.

grid An open frame of timber beams resting on the foreshore which supports a vessel being repaired during low tide. See SLIP (2).

grip A shallow gutter formed along the margin of a road to carry rainwater from the road to a drain or ditch. See CATCHWATER DRAIN.

grit chamber A small DETRITUS CHAMBER for collection and removal of solids.

groins (U.S.A.) See GROYNES.

gross capacity (reservoir or tank) Volume of water stored between bed level and the sill level of the waste weir.

gross duty of water (or **diversion requirement**) The water diverted at the intake of a canal system and used for irrigation (U.S.A.). See NET DUTY (of water).

ground moraine The debris which accumulates in the sole of a glacier and is steadily pushed or carried forward with the ice; the debris deposited by a glacier when the ice melts. See ENGLACIAL MATERIAL.

110

groundwater Water in the ZONE OF SATURATION; SUBSURFACE WATER as opposed to SURFACE WATER; groundwater has a number of advantages over surface water as supplies, e.g. better bacteriological and chemical quality, greater scope for well siting, and, commonly, lower cost of schemes.

groundwater artery A channel or passage saturated with ARTESIAN WATER and surrounded by a CONFINING BED. A groundwater artery is often a BURIED CHANNEL or depression filled with saturated gravel or other permeable material.

groundwater basin A basin-shaped group of rocks containing groundwater with geologic and hydraulic boundaries convenient for investigation and description. The basin commonly embraces both the recharge and discharge areas. See GROUNDWATER PROVINCE.

groundwater budget A statement or estimate of water resources; usually applied to a groundwater basin or province; it gives estimates of storage, discharge, and recharge of water and other assessments and forecasts. See WATER CONSUMPTION, WATER RESOURCES.

groundwater dam A subsurface mass or barrier (such as a fault or dyke) which prevents or impedes the lateral movement of groundwater and results in a marked difference in level of the water-table on the two sides of the dam.

groundwater decrement (or **phreatic water discharge**) A decrease in groundwater storage by withdrawal from wells, spring flows, infiltration tunnels, and evaporation discharge.

groundwater depletion curve See DEPLETION CURVE, NORMAL DEPLETION CURVE.

groundwater discharge Discharge of water from ZONE OF SATURATION into bodies of surface water or upon land. See EFFLUENT SEEPAGE, SPRING.

groundwater divide (or **phreatic divide**) The line of maximum elevation along a groundwater ridge where the water-table slopes downwards in opposite directions (Figure G.1). See DIVIDE.

groundwater equation The balance between water supplied to a basin and the amount leaving the basin. The information requires the collection of comprehensive data of water intake and discharge.

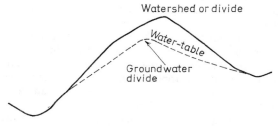

Figure G.1. Groundwater divide

111

groundwater flow   Part of stream flow derived from zone of saturation through seepage or springs. See EFFLUENT SEEPAGE.

groundwater increment (or **groundwater recharge** or **recharge of aquifer**)   Broadly, the recharge or replenishment of water in the zone of saturation; may be expressed in cubic metres, gallons, or acre feet.

groundwater inventory   A detailed statement or estimate of quantities of water forming GROUNDWATER INCREMENT balanced against estimates of quantities forming GROUNDWATER DECREMENT for a particular area or basin. See GROUNDWATER BUDGET.

groundwater lowering   Lowering the water-table locally so that excavations may be made in relatively dry conditions; it also helps to strengthen the ground excavation sides. The water may be lowered by WELLPOINTS located along the fringe of the excavation, or by FILTER WELLS, which are preferred for deep lowering. The effective limit of a wellpoint is about 5·5 m (18 ft). See ARTESIAN WELL (DRAINAGE).

groundwater mound   A ridge or elevation formed in a body of groundwater by INFLUENT SEEPAGE.

groundwater provinces   Districts, areas, or basins throughout each of which the groundwater conditions are similar.

groundwater recession   The general sinking or lowering of the water-table of a basin or area.

groundwater recharge   See GROUNDWATER INCREMENT, RECHARGE OF AQUIFER.

groundwater runoff   Runoff which existed, wholly or partly, as groundwater since its last precipitation. See DIRECT RUNOFF, SHALLOW PERCOLATION.

groundwater storage   Estimate of quantity of water in ZONE OF SATURATION; that stage of the HYDROLOGIC CYCLE when water is leaving and entering groundwater storage.

groundwater storage curve   A curve showing quantity of groundwater available for runoff at given rates of GROUNDWATER FLOW. See NORMAL DEPLETION CURVE.

groundwater tracers   Dyes, chemicals, or salts mixed with surface waters to trace the source of water seeping into tunnels, wells, shafts, or deep excavations. Some radioactive isotopes form good tracers because of the high sensitivity with which they can be detected after seeping from surface or other source to point of issue underground. Tritium, a heavy isotope of hydrogen, has outstanding tracer qualities. See DARCY'S FORMULA.

groundwater trench   A relatively narrow depression in the water-table caused by EFFLUENT SEEPAGE into a stream, channel, or drainage ditch.

group action   The planning, application, and maintenance of land, soil, and water conservation measures by a group of ranchers or farmers working with a leader or technician (U.S.A.).

group enterprise   The design and installation of a soil and water conservation scheme, which serves more than one ranch or farm, through the joint effort of those served. The scheme may involve the construction of a dam, canal, or main drainage channel or enlargement of the irrigation system.

grout curtain   See CURTAIN GROUTING.

grouting   (1) The injection, by pumps through boreholes, of cement slurry or chemicals into the ground around tunnels, excavations, shafts, or dam foundations. When set, the grout reduces or prevents the inflow of water and also improves the mass strength and elastic properties of the rocks. Cement may be used to fill the larger fissures and followed with chemical grout to seal the smaller cracks and voids. (2) The grouting of banks to form resistant masses to check the lateral erosion of important land areas. See BLANKET GROUTING, CHEMICAL CONSOLIDATION, CLAY SEALING, CURTAIN GROUTING.

grouting (boreholes)   The grouting of the annular space between the ground and the lining of the boreholes. It is often carried out for the first 30 m (100 ft) or so, mainly to retain the lining in position, to exclude unwanted water, and to support loose ground. The grouting is performed by pipes from the surface or from the bottom upwards.

groyne, filling for   Rounded stone, large gravel, or shingle is generally used, depending mainly on the type of groyne and river conditions. Earth, sands, silts, and small gravel are unsuitable as the material tends to be washed out through the groyne structure by the water.

groynes (or **groins** (U.S.A.))   (1) River (Figure G.2). Elongated structures with one end, or root, in the bank of a river and the other end, the head or nose, projecting out into the channel. They may be impermeable, semipermeable, or permeable. The construction material may be temporary (weighed FASCINES, etc.), semipermanent (piles and sheeting or timber sheet piling), or permanent (concrete, masonry, etc.). They are erected at right angles to the bank or current or given an upstream or downstream slope. It is usual to use groynes in series so that the effect is continuous.

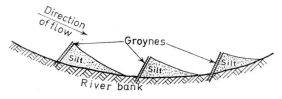

Figure G.2.   Groynes

A series of river groynes will check erosion and induce the deposition of silt in the sluggish areas between the groynes. (2) Coast. Usually timber, but sometimes stone or concrete structures erected normal to the coastline or to the prevailing LONG-SHORE DRIFT, to accumulate or retain beach deposits where there is LITTORAL DRIFT. In general, groynes are set at a distance apart about equal to their length, though this varies with local conditions. They are fixed sufficiently deep to avoid scour at low tides. See A-FRAME GROYNES, CRIB GROYNES, DIRECTION OF GROYNES, FASCINE GROYNES, SINKER GROYNES, STONEMESH GROYNES, WILLOW GROYNES.

groynes, auxiliary protection   See AUXILIARY PROTECTION (GROYNES).

groynework   All works and practices in which groynes are used to control and protect rivers, coasts, and banks. It includes the planning, design, location, construction, and protection of groynes.

guard lock   A lock which keeps a dock separated from tidal waters.

guard post   See BOLLARD.

gulley   A pit, chamber, or receptacle of concrete, stoneware, cast-iron, or other material, covered with a grating and placed on a roadside to collect water from a gutter or channel. See GULLEY SUCKER, YARD GULLEY.

gulley sucker   A tanker lorry for removing silt and mud from road gulleys; a pump sucks out the material and forces it into the tank. See YARD GULLEY.

gulley trap (or yard trap)   A water seal fitted in a gulley to prevent foul gases escaping from the drain.

gully   A ditch, gutter, or channel cut by water and along which water usually flows only after heavy rainfall or melting of snow or ice.

gully control plantings   To establish or re-establish a vegetative cover sufficient to control runoff and erosion by planting seeds, cuttings, or transplants in gullies. See COPPICING, TREE PLANTING.

gully dam   A small FARM DAM constructed across a drainage depression or gully for the purpose of impounding water for irrigation or other farm purposes. See HILLSIDE DAM.

gully drainage (or gully erosion)   Drainage which is concentrated in gutters or furrows. Gullying leads to rapid drainage and erosion of valuable soil may also be rapid if vegetation is sparse or has been removed. See SHEET EROSION.

gully erosion   See GULLY DRAINAGE.

gully stabilisation structure (or waterway stabilisation structure)   A structure designed to prevent or minimise bed scour or other erosion in a channel or gully. The structure does not provide floodwater storage.

guttation   Dropping or flowing in drop-like masses; the expulsion of water, in excess of transpiration, from uninjured leaves of vegetation.

gutter   Any passage for water; a pipe or tube for conveying water from the roof or eaves of a building; a ditch or channel in the ground and lined with CLAY PUDDLE for conveying water alongside a canal, road, or street, to form into small longitudinal channels or gullies, usually by rain water.

H

hailstorm A STORM, often prolonged and severe, consisting largely of hail or frozen raindrops ranging in diameter from 5 mm to 10 mm or more; the ice particles are precipitated either separately or as aggregates of irregular size and shape. See THUNDERSTORM, SLEET, SNOW.

half-socket pipe An AGRICULTURAL DRAIN of which only the lower half is socketed. If made of concrete, the lower half may be impervious and the upper half porous.

half-tide cofferdam A COFFERDAM in an estuary or the sea in which the height is not sufficient to exclude water at high tide, so that after every full tide de-watering is necessary. See FULL-TIDE COFFERDAM.

hand boring See AUGERS.

hand clean-up-dam foundation A term used in some earthwork contracts in which unit rate includes hand excavation and loading, cleaning by air and water jets, but not guniting, slush grouting, etc.

hand lead Sometimes applied to a SOUNDING WEIGHT.

hand level See ABNEY LEVEL, BONING, CLINOMETER.

hanging glaciers See CIRQUE GLACIERS, GLACIERETS.

hanging valley A tributary valley which is discordant as to grade with its main valley. May have been formed during the glacial period when the main valley was deepened by glacial erosion at a faster rate than that of its tributary, which formed a hanging valley when the ice disappeared. The considerable fall of water at the mouth of these valleys is sometimes used for power generation.

harbour A natural or artificially sheltered area of water where ships can anchor; it may have facilities for loading and discharging. See ARTIFICIAL HARBOUR, BREAKWATER.

harbour models Models of proposed harbours constructed to a scale ranging from 1 in 50 to 1 in 180, with outside waves of about 20 mm ($\frac{3}{4}$ in) high. Harbour models are useful during planning and design and they also assist in solving problems of scour and silting. See MODEL ANALYSIS.

harbour of refuge A harbour which only gives ships shelter during storms and has no loading or unloading facilities.

hardness of water Refers to the presence of dissolved impurities in water. A hard water does not lather readily when mixed with soap, as pure or soft water does. The main salts in solution which cause hardness are the carbonates and sulphates of calcium and magnesium. TEMPORARY HARDNESS (or CARBONATE HARDNESS) produced by the presence of the carbonates can be removed by boiling or by other means; PERMANENT HARDNESS (or NON-CARBONATE HARDNESS) produced by the presence of the sulphates can be removed by the addition of washing soda or by chemical treatment. Water hardness is expressed in parts per million (by weight) or milligrammes per litre. See BRACKISH, FRESH.

hardpan (or **pan**) A hard layer sometimes found below the topsoil which is relatively impermeable to water and does not soften when wet. The grains have been cemented together by iron or calcium salts leached from the topsoil. See BOUND GRAVEL.

Hawksley's formula A formula used in the U.S.A. and U.K. to determine the height of waves in reservoirs:

$$H = 0{\cdot}025\,\sqrt{L}$$

where H is the height of wave in feet and L is the FETCH in feet.

Hawksley, Thomas (1807–93) A distinguished British water engineer who was responsible, together with John Frederic La Trobe BATEMAN, for most of the major waterworks constructed in the U.K. between 1830 and 1890.

haze See FOG.

Hazen's law A formula giving the approximate permeability of soils based on their effective grain size, and equals D10 size $\times$ 100 in cm/s. See EFFECTIVE SIZE, GRADED FILTER.

head The potential energy of water due to its height above a given datum; usually given in feet. See ARTESIAN HEAD, FRICTION HEAD, HYDROSTATIC LEVEL, HYDROSTATIC PRESSURE, PRESSURE HEAD, VELOCITY HEAD.

head bay The part of a canal lock immediately upstream of the lock gates.

head-control gate A type of gate often used in CHANNEL CHECKS to regulate the water level. The gate is set to act as an adjustable-level weir.

head, elevation See POTENTIAL HEAD.

head, entry See ENTRANCE HEAD.

header (1) A conduit or pipe which conveys water to or from another conduit or pipe system. (2) A stone or brick laid at right angles to the face of wall.

head flume A flume placed at a junction with a terrace outlet to prevent cutting, or at the head end of a gully to prevent the extension of the gully.

headgate (or **head gate**) The upstream gate or control device through which water may enter a conduit. See TAILGATE.

headings (wells) Small tunnels, adits, or headings excavated into the water-bearing rock formations to increase the yield of a well. The scheme may include (1) drilling large-diameter holes from the surface to intercept the headings; enlarging fissures in the headings, (2) drilling small-diameter holes in the roof, sides, and floor of the headings. See ADIT, INFILTRATION GALLERY.

heading up The upstream rise in water level from a construction because of regulation of flow as when gates at a barrage are closed. See HYDRAULIC BORE.

head, kinetic See VELOCITY HEAD.

head loss (or **lost head**) See FRICTION HEAD, HYDRAULIC FRICTION.

head of groyne See GROYNES.

head race A channel along which water flows to a turbine from a FORE-BAY.

head wall   A retaining wall, usually concrete, placed at the inlet and outlet of a culvert, drain, or other hydraulic structure on water supply channels (Figure S.2).

headward erosion   The erosion of the steepening hill slopes and the gradual extension of valleys and streams; the upstream movement of a scour or waterfall. Where necessary, the development of a deep scour or waterfall may be arrested by the use of heavy stone, stonemesh sills, or a concrete or timber sill. See CHECKDAM.

headwater control   Flood control measures in a headwater drainage area. See FLOOD CONTROL.

headwaters   The waters upstream of a structure; the upper reaches or source of a stream. See TAIL WATER.

headworks (or **intake heading** (U.S.A.))   The diversion structure at the head of main irrigation canals and water supply conduits.

heave   See FROST HEAVE, SWELL (2).

heavily brackish   See BRACKISH.

heel   The part of the base on the earth side of a retaining wall, or the part of the base on the water or upstream side of a dam. See TOE.

heeling   The traditional method of mixing wet clay and sand to make a water-tight canal bed. The kneading of this mixture is done by labourers who 'heel' it with their boots. Developed by James Brindley in the 18th century and still used today on certain works. See PUDDLING.

heel post (or **quoin post**)   A corner post of a lock gate. See HOLLOW QUOIN.

height of capillary rise   See CAPILLARY RISE.

height of weir   The height from bed to crest on the upstream side.

held water   A term sometimes applied to CAPILLARY WATER; water retained in the ground above the STANDING WATER LEVEL.

helium diving bell   A DIVING BELL in which a helium–oxygen mixture is breathed by the men instead of ordinary air.

hencooping   A form of CRIBWORK consisting of triangular cribs (or hencoops) filled with stone and planted with willow spars which grow and hold the cribs in position along the river bank.

heeringbone drain   See CHEVRON DRAIN.

high-alumina cement   A cement with the property of rapid hardening; prepared by fusing calcareous and aluminous materials to the completely molten state and crushing the resulting clinker to a fine powder. See PORTLAND CEMENT.

high bottom phase   See FIRST BOTTOM.

higher critical velocity   That velocity when eddies are first noted; LOWER CRITICAL VELOCITY is that at which eddies in originally TURBULENT FLOW die out.

higher high water (H.H.W.)   The higher of the two high waters of any lunar day; when tide is diurnal, the high water occurring daily is accepted as higher high water.

higher low water (H.L.W.)   The higher of the two low waters of any lunar day.

117

highest high water (H.H.W.) The highest known water level.

high water (H.W.) The highest level reached by a rising tide; the highest water level within any given period.

high waterline (tide) The line where the plane of mean high tidal water intersects the shore.

high water mark The highest level reached by water during a flood as marked by debris or silt.

highway erosion control Includes engineering and vegetative practices for the prevention and control of erosion at cross drains, in ditches, and on fills and road banks within a highway. See GUTTER.

hillside dam A small FARM DAM, usually curved or three-sided, built on the sloping side of a hill with no well-defined gully or depression. This type of dam may have STORAGE RATIOS up to 3 : 1. See GULLY DAM, RING DAM.

hillside swamps Swamps on mountain sides formed by a local discharge of subsurface water, or by depressions or the downward percolation of moisture from melting snow. See COASTAL PLAIN SWAMPS.

histogram (or **frequency diagram**) A diagram or curve showing frequency distribution and so constructed that the area below the curve corresponds to the frequency. See FREQUENCY CURVE.

historic flood A flood whose date is known, and possibly its magnitude, but which occurred before the establishment of regular river gauging or recordings. See FLOOD (RETURN PERIOD).

hollow dam A dam built of reinforced concrete, plain concrete, or masonry, in which the water pressure is taken on an inclined slab or vault carried by buttresses spaced at regular intervals. See MULTIPLE-ARCH-TYPE DAM.

hollow quoin Recessed masonry with a drilled hole in its upper surface carrying the HEEL POST of a LOCK GATE.

homogeneous dam Applied to an earth dam built more or less from one material (Figure O.1).

hooked spits Spits terminating in a curve; formed by strong currents flowing past their ends.

hook gauge An appliance for measuring the elevation of the free surface of a liquid; consists of a pointed hook attached to a vernier which slides along a graduated staff. The hook is lowered into the water and raised until the upward point just cuts the water surface. It measures water level with great accuracy. See POINT GAUGE.

hook groyne A type developed in New Zealand, consisting of a bank of shingle, curved in plan, with an armoured head pointing upstream. Above and below the groyne, heavy willow planting traps the silt, particularly on the upstream side.

hopper barge (or **dumb barge** or **scow**) A barge for transporting dredged material from a DREDGER to a site for dumping. It may be equipped with power and discharging facilities. See FLOATING PIPELINE.

hopper dredger   A DREDGER built to contain and carry its dredged material to the dumping ground.

horizontal alignment   See REALIGNMENT.

horizontal drainage blanket   A DOWNSTREAM DRAIN frequently placed in moderately high earth dams. It is located in the base of the dam and covers about a third of the downstream area. It may consist of 0·3 m (12 in) compacted layers of cobbles, gravel, and sand built up to a thickness varying between 1·5 and 3·0 m (5 and 10 ft).

hot springs   Springs with a constant discharge of hot water which is sometimes used for domestic purposes as in Iceland. See GEYSER.

house drain (or **collection line** (U.S.A.))   A drain or drains taking all sewage or wastes from a house. In the U.K. it extends from the dwelling to the point of connection with the local authority's public sewer. See COLLECTING SYSTEM, COMBINED SYSTEM, SEPARATE SYSTEM.

humid   Damp, moist; climate or land where agriculture can be carried out without irrigation because the moisture, when distributed normally throughout the year, is adequate.

humidity   A measure of the water vapour content of the atmosphere; may be expressed as RELATIVE HUMIDITY or as ABSOLUTE HUMIDITY or the mass of water present in a cubic metre of the air. See HYGROMETER, SATURATION DEFICIT.

humidity, absolute   See HUMIDITY.

humidity, relative   See RELATIVE HUMIDITY.

humid, semi-   See SEMI-HUMID.

humus tank   The final settling tank from which sewage effluent passes out to the land or into a stream.

hurdle groynes   Used for checking erosion; consist of light, driven piles with saplings secured horizontally to them and also brushwood occasionally. The groynes are placed about a chain apart along the bank (Western Australia).

hurdle work (or **wattle work**)   A low fence on a river bank, consisting of osiers (species of willow) interlaced with vertical sticks, to retard scour and encourage silting. Often cheap and effective along smaller streams with a sandy bed, as well as along the margins of lakes to check erosion by wave action. Has been used successfully in Victoria, Australia.

Hutt River Reclamation   A scheme by which the Hutt River Board of New Zealand reclaimed shingle flats of the Hutt River and converted them to good grassland. It involved the building of small, low stonemesh weirs across the flats at right angles to the riverflow. They were so spaced that the shingle caught on the upstream of one extended to the apron on the next above. At this stage, another weir was erected upstream and so on until a long stretch of unproductive river shingle was reclaimed. See LAND ACCRETION.

hydrant (irrigation)   A small box or chamber containing a valve and coupling, with a removable cover flush with the surface; used for connecting irrigation spraylines or sprinkler laterals to MAIN PIPES. Any pipe with nozzle at which water may be drawn by means of a hose. See STAND-PIPE.

hydraulic Relating to the flow or conveyance of liquids, especially water, through pipes or channels; the use of water as motive power. See HYDRAULICS, HYDRODYNAMICS.

hydraulic bore A wave in an open channel which advances upstream from a point where flow has stopped or diminished because of obstruction. See SURGE, TIDAL CURRENT.

hydraulic cement A special type of cement which possesses the property of hardening or setting under water.

hydraulic check nut A fitting used for the protection of pressure gauges against sudden pressure releases.

hydraulic conductivity Sometimes used for COEFFICIENT OF PERMEABILITY.

hydraulic discharge The loss of groundwater by discharge through springs, pumping from wells, etc. See GROUNDWATER DECREMENT.

hydraulic dredger See SUCTION DREDGER.

hydraulic ejector (or **elephant's trunk** or **silt ejector**) A pipe arrangement for removing mud, sand, or silt from the chamber at the bottom of a pneumatic caisson. Water is injected into the bottom end of the pipe by means of a pump at the surface. The removal of mud and silt depends on the ejection principle.

hydraulic elements The values that affect the flow of water in a conduit, such as velocity, area, depth, wetted perimeter, and so on.

hydraulic elevator An arrangement widely used in various goldfields, towards the end of the 19th century, for lifting gravel and sand up to the drainage or discharge level. A jet of water creates a powerful suction in a hopper and the gravel and water are carried up a pipeline and then discharged along sluice-boxes. See HYDRAULIC EJECTOR.

hydraulic excavation The use of powerful water jets to break down deposits of coal or gravel containing gold, tin, etc. See HYDRAULIC MINING.

hydraulic extraction (or **hydro-extraction**) A general term for the various processes of excavating and transporting coal, metals, minerals, and gravels by water power.

hydraulic fill The filling of waste spaces, embankments, or construction cavities by sediment or debris carried by water through a pipe or a flume. The filling of waste spaces in mines by waterborne material is known as hydraulic stowing.

hydraulic fill dam An earth dam built up from gravel and silt carried into place by pipelines or sluiceways located along the outer slopes of the dam, with the waterborne material flowing inwards towards midstream. The coarse material settles on the outer shell (towards the banks) of the dam, while the fines are carried into the central part to form a more or less impervious core or core wall (Figure H.1).

hydraulic flume transport The use of flowing water to transport pulp, broken coal, mineral, or gravel in rectangular or semicircular channels.

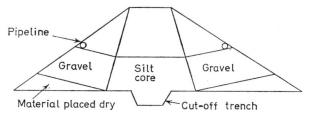

Figure H.1. Hydraulic fill dam

The channel slope should not be less than 3°. Movement of broken coal in flumes commences at a water velocity of about 1 m/s (3 ft/s), but in practice velocities of at least 2 m/s (6 ft/s) are used.

hydraulic friction (or **loss of head**) The friction or resistance to flow of the wetted surface of a channel, conduit, or pipe; eddies and cross currents normally associated with TURBULENT FLOW are included. Whenever possible, the effect of obstructions, excessive channel variations, curvatures, impacts, etc., are not included in hydraulic friction. See ROUGHNESS COEFFICIENT.

hydraulic grade line Used in U.S.A. for HYDRAULIC GRADIENT.

hydraulic gradient In a closed conduit, an imaginary line connecting the points to which water will rise in vertical open pipes extending upwards from the conduit; in an open channel, it is the free surface of flowing water. The gradient along a closed conduit may be expressed as the drop in feet per mile or metres per kilometre or as a ratio.

hydraulic jump A rapid and often turbulent flow of water from low stage below CRITICAL DEPTH to high stage above critical depth; the CRITICAL VELOCITY changes from SUPER-CRITICAL FLOW to SUB-CRITICAL FLOW (Figure H.2).

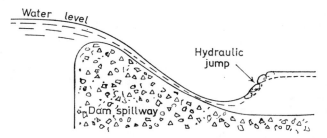

Figure H.2. Hydraulic jump

hydraulicking The use of a strong jet of water from a nozzle to break down and remove a face of alluvial wash or other deposit and the subsequent recovery of the gold or other valuable mineral so set free in sluice boxes

121

or other appliances. The method is adopted only in undeveloped regions and where water is plentiful.

hydraulic loading The sluicing or flushing of broken material, obtained by hydraulic excavation, along the floor and into FLUMES or other appliances convenient for loading the material. In the case of broken coal, the material flows towards the flume if sufficient water is available and the floor slope is not less than 6–7° in favour of the flow. The flushing operation may be accelerated by flexible hoses with 1·05 to 1·4 N/mm² (150–200 lbf/in²) water pressure.

hydraulic main A main pipeline which supplies water under high pressure to subscribers for driving hydraulic machines, such as lifts, cranes, and so on. See MAIN PIPES (IRRIGATION).

hydraulic mean depth See HYDRAULIC RADIUS.

hydraulic mining (1) Surface. The use of high-pressure jets of water for breaking down relatively soft deposits of gravel or sand containing gold or tin or other valuable material. The washed debris flows along flumes with ribs or riffles in the bottom which trap the gold or other heavy mineral. (2) Underground. Applied mainly to the excavation and extraction of coal by high-pressure water jets. The jets mpel the coal fragments along the floor to the point of collection. Tests indicate that water at a pressure of about 4·2 N/mm² (600 lbf/in²) may be adequate to break down the softer coals, while pressures up to 8·4 N/mm² (1200 lbf/in²) may be required for the harder coals. In the harder coals it may be necessary to pre-crack or loosen the seam by shotfiring or water infusion before applying hydraulic mining.

hydraulic models A scale representation of a hydraulic structure which is geometrically similar at all solid–liquid boundaries; a close general resemblance to its prototype is also valuable. The type of flow in the model must be the same as in the full-size plant. Model experiments ensure considerable accuracy in the solution of many hydraulic problems. See FROUDE NUMBER, HARBOUR MODELS, MODEL ANALYSIS.

hydraulic permeability The capacity of a rock or soil for transmitting water under pressure; may be different in different directions.

hydraulic press See HYDROSTATIC PRESS.

hydraulic pressure snubber A SNUBBER used in self-contained hydraulic units to protect pressure gauges against sudden fluctuation of pressure and pressure ripple.

hydraulic profile (aquifer) A vertical section of the PIEZOMETRIC surface, or the surface to which water will rise under its full head, from a given aquifer.

hydraulic radius (or **hydraulic mean depth**) The cross-sectional area of a stream divided by its WETTED PERIMETER; used in formula to predict the velocity of water flow in an open channel. For a given channel and slope, the greatest hydraulic radius gives the largest flow.

hydraulic ram pump A self-acting pump which uses the momentum of a fall of water in a stream to force a small quantity of water to an elevation considerably higher than the initial stream fall. If correctly installed it will operate for many years without maintenance.

hydraulics The engineering application of the principles of HYDRODYNA-MICS. In mining, the applications include wet coal or ore treatment plants, drilling fluids, water infusion, hydraulic stowing, and hydraulic power. In civil engineering, it is applied to the flow of water in rivers, open channels, irrigation, drainage, seepage, dam construction, and water supply.

hydraulic stripping The removal of overburden or waste rock overlying an economic deposit at the surface by hydraulic methods.

hydraulic test (1) A water test to locate any leakage of water in newly laid drains. They are closed at the lower end and then filled with water and kept to a maximum head of 2·1 m (7 ft) for an hour. The test is satisfactory if no fall of level occurs during the hour. (2) A general test for pipes, pressure vessels, boilers, etc., which are filled with water and subjected to the designed pressure or slightly in excess.

hydraulic transport The conveyance of broken or crushed material by means of water flowing in flumes or pipes.

hydrodynamics The scientific study of the flow of fluids, particularly water flowing through pipes, in open channels and orifices and over weirs and notches, including the problems relating to energy and pressure. See HYDROSTATICS.

hydroelectric power Electrical power obtained from a water-driven dynamo. See HYDROELECTRIC POWER STATION.

hydroelectric power station A building in which turbines are rotated, to drive generators, by the energy of natural or artificial waterfalls. See PUMPED STORAGE, PUMPED-STORAGE HYDRO STATION.

hydroelectric scheme A major and complete project for using water power to generate electrical energy; scheme usually includes dams, spillways, tunnels, buildings, roads, bridges, housing accomodation, and many other works, services, and facilities. See DAM, RESERVOIR.

hydro-extraction See HYDRAULIC EXTRACTION.

hydrogenesis The natural condensation of moisture in the air voids in soils and surface deposits.

hydrogeological map A map which depicts, from observed facts, the surface distribution of water and the principal aquifers, together with their thickness, lithology, and outcrop belts in the area covered. It may also show structure contours on the aquifers and perhaps the ISOPACHYTES of the formations. The sites of the major abstraction wells and all surface springs and water supplies are shown. It may include sections across important aquifers and catchment areas. Most of the hydrogeological maps now being published are scaled at between 1 : 50 000 and 1 : 200 000.

hydrogeology The science concerned with the occurrence and movement of both surface water and groundwater; includes a considerable amount of geologic orientation. See GEOHYDROLOGY, HYDROLOGY.

hydrograph A graph showing level, velocity, or discharge of water in a channel or conduit plotted against time. The most common graph is discharge against time, with time expressed in hours, days, weeks, or months.

hydrograph, flood See FLOOD HYDROGRAPH.

hydrographer One who surveys and draws maps of the seas, lakes, rivers, and other waters including the adjacent shores; one who measures and records rainfall, runoff, water level, water flow, and associated quantities.

hydrographic datum A datum used as a reference for measurements of heights of tides, levels of rivers or canals, or depths of water in wells and underground storage. See ORDNANCE DATUM.

hydrograph, recession (or **normal recession curve**) The curve obtained from specified lengths of hydrograph that represent discharge from channel storage or a natural valley after subtracting BASE FLOW; the curve showing the decreasing rate of flow in a stream channel.

hydrograph, well See WELL HYDROGRAPH.

hydrography The study and measurement of seas, lakes, rivers, and other waters including their marginal land areas, and the preparation of maps, charts, and publications for use of navigators in such waters. See OCEANO-GRAPHY.

hydroisobaths Lines connecting points of identical depths of water-table below ground surface. See WATER-TABLE CONTOUR PLAN.

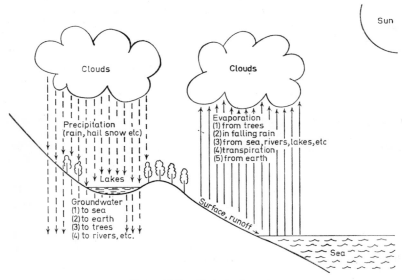

Figure H.3. Hydrologic cycle

124

hydroisopleth map A map showing fluctuations in position of water-table with respect to time.

hydrologic cycle The series of transformations in the circulation of surface waters to atmosphere, to ground as precipitation, and back to surface and subsurface waters (Figure H.3).

hydrologic processes Processes which may be divided into two main phases: (1) the wetting phase, when precipitation is the dominant process, (2) the drying phase, when evaporation is the dominant process with no precipitation. During the wetting phase, evaporation diminishes and the catchment gains water from rainfall through infiltration and depression storage. See MOISTURE STATUS.

hydrology The science concerned with the occurrence of water on and in the earth, its physical and chemical properties, and the changes involved from its period of precipitation until its discharge into a sea or other waters or return to atmosphere, that is, during a complete HYDROLOGIC CYCLE.

hydro-mechanisation The use of modern hydraulic methods of excavating and moving coal or other minerals in surface or underground operations.

hydrometeorology The study of METEOROLOGY in relation to hydrology.

hydrometer An instrument for measuring the density or specific gravity of liquids. The ordinary type consists of a slender graduated stem fixed to a weighed bulb; when floating vertically in the liquid to be rested, the specific gravity is indicated on the stem. In liquids of low density a shorter length of stem is exposed than in liquids of high density.

hydrometry Concerned with, or determination of, the SPECIFIC GRAVITY or density of liquids. Sometimes used in relation to the measurement of velocity or discharge of waters as from reservoirs, rivers, and other waters.

hydrosphere The waters of the earth's crust; lying partly within as groundwater and partly above as oceans and other waters and also as moisture in the atmosphere.

hydrostatic joint A spigot-and-socket joint in a water main. An hydraulic ram is used to force lead into the socket.

hydrostatic level (or static level) The level to which water will rise in a tube, pipe, or well under its full PRESSURE HEAD.

hydrostatic press (or hydraulic press) An appliance, based on PASCAL'S LAW, in which a force applied by a piston over a small area is transmitted through water to another piston having a large area; by manipulating the small piston high pressures may be obtained.

hydrostatic pressure The pressure at any given point in a liquid at rest; equals its density multiplied by the depth. See PRESSURE HEAD.

hydrostatics The scientific study of the properties of, and particularly the pressures and forces in, water at rest.

hydrostatic test See HYDRAULIC TEST.

hyetal Relating to rain, or to its distribution. See ISOHYETALS.

hyetal interval The rainfall difference represented by two ISOHYETALS.

hyetal regions Major world divisions based on rainfall characteristics.

hyetograph A chart or graph showing average rainfall or average intensity of rainfall; a recording rain-gauge.

hygrograph A self-recording HYGROMETER.

hygrometer (or psychrometer) An instrument for measuring the RELATIVE HUMIDITY of air. In the whirling type, two thermometers (dry-bulb thermometer and wet-bulb thermometer) are fixed side by side on a brass frame and fitted with a loose handle so that it can be whirled in the air to be tested. After whirling for about one minute at approximately 200 revolutions per minute, the readings on both thermometers are recorded and by referring to Glaishers or Marvin's hygrometrical tables a consistent and precise figure can be obtained for the relative humidity.

hygroscopic coefficient The amount of moisture a dry soil will absorb in nearly saturated air (98% RELATIVE HUMIDITY) at 25 °C (77 °F). The result is given as a percentage of the dry weight of the soil.

hygroscopic moisture The moisture contained in air-dried soil but which is driven off if the soil is dried at 105°C.

hyper-critical flow (or super-critical flow) A flow in which the velocity is greater than one of the recognised values of CRITICAL FLOW. See SUB-CRITICAL FLOW.

hypogene Changes, processes, or forces within the earth's crust, such as earth movements.

I

ice Water in the solid state. Water freezes at O °C and expands and becomes less dense than water. See FROST ACTION, GLACIER, SNOW.

ice age The Pleistocene geological period, covering some hundreds of thousands of years and ending only a few thousand years ago. The glaciers and ice sheets covered large continental areas, including at times nearly all Britain north of Bristol Channel and the Thames Valley. Extensive deposits of ice-borne material were left and around some coasts new land was added; many of the countries affected by the glacial period have published papers on these deposits which are of value to water engineers as they give information as to thickness of drift and presence of BURIED CHANNELS, etc.

ice apron (or ice breaker) A structure for breaking ice and thus protecting a bridge pier; consists of a ramp on the upstream side of a bridge and sloping upwards from below water level.

iceberg A huge floating mass of ice. When a glacier reaches the sea, it begins to float and break up into icebergs of varying shape and size.

ice caps Thick and widespread sheets of glacier ice in regions where the SNOW LINE is low.

ice cover A mantle of ice over an area of land or water and particularly its thickness.

ice gorge (or **ice jam**) The piling up of ice against a restriction or obstruction in the channel of a stream and forming a temporary bar or dam. The sudden release of water when the obstruction collapses often causes flooding of any low-lying areas in the valley.

ice-pushed terrace A terrace formed along some lake shores by the thrust of ice up the beach. In northern latitudes, similar terraces have been noted on the shores of fiords and inlets of the sea.

ice ramparts (lakes) Terraces or ridges of gravelly material along the shores of lakes; caused by the contraction and expansion of the ice cover during cold weather. Sometimes the ice thrust causes damage to structures along the shore of such lakes.

imbibition Absorption of water by plants from the soil; absorption of water into the pores of a rock or soil.

Imhoff tank A two-storeyed tank for the fermentation of sewage; methane gas is formed and the sludge which settles out is collected and dried.

immediate runoff See SURFACE RUNOFF.

impact loss The energy or head lost by impact or collision of particles of water; included in EDDY LOSS.

impact pressure The pressure exerted during impact or collision between a solid and a fluid.

impeller The rotating curved blades of a CENTRIFUGAL PUMP or TURBINE PUMP. See DIFFUSER CHAMBER.

impeller pumps See CENTRIFUGAL PUMP, SUBMERSIBLE PUMP, TURBINE PUMP.

impermeability factor (or **runoff coefficient**) A factor which enables runoff to be calculated, and is the ratio of the DIRECT RUNOFF to the average rainfall over the whole drainage area for any storm.

impermeable (or **impervious**) Describes a soil, rock, or other substance which permits the passage of water only at a very slow rate. See HAZEN'S LAW, DARCY'S FORMULA.

impermeable groynes A type of GROYNE often used for confining, straightening, or narrowing a river; it induces scouring along specific lines to eliminate shallow, wide, and meandering lengths. These groynes are usually long and to keep down costs are often made of earth or shingle with a head reinforced with stonemesh crates or concrete blocks. Their crests are built well above flood level to avoid overtopping and erosion. Each groyne is built at one time. Many of the river control groynes used in New Zealand are of this general type and known as ASHLEY TYPE GROYNES. See PERMEABLE GROYNES.

impervious rock A rock that will not permit the passage of water, oil, or natural gas, or only at a very slow rate. Impermeable rocks include shales, clays, compact limestones, marls, and unfissured igneous rocks. See AQUIFUGE.

impounding reservoir A large water storage with sufficient capacity to satisfy a water supply scheme over a considerable period of time. See RESERVOIR, RESERVOIR BASIN, SERVICE RESERVOIR.

impounding schemes (planning) Field investigations and research so that the impounding reservoirs make full use of the catchment area. In planning, it may be decided (1) to construct a single dam to the full height required to develop the catchment. This is often adopted where only one site is favourable, or (2) to construct the first dam in such a position to enable further dams being built as later stages of the entire scheme. In general, it is considered advantageous to exploit the possibilities of a single dam site to the maximum. See DAM SITE INVESTIGATION, WATERWORKS PLANNING.

impregnation Any method which prevents or reduces water seepage through ground by saturating it with a waterproofing liquid. See GROUTING.

improved Venturi flume An earlier term for PARSHALL MEASURING FLUME.

impulse turbine A WATER TURBINE (e.g. the PELTON WATER WHEEL) which acquires its rotative force from the impulse of a jet or body of water rather than by a fall in its pressure. See REACTION TURBINE.

inches of water Storm runoff is often expressed as inches of depth uniformly covering the given drainage area. One inch = 25·4 mm.

incised meanders A feature formed when a river, after it had reached a meandering stage, is rejuvenated and cuts downwards once more, and so carves a gorge-like valley along the old meander channel. See VALLEY-IN-VALLEY STRUCTURE.

incised river A river whose channel has been formed by DEGRADATION. See ALLUVIAL RIVER, SUBSEQUENT STREAM.

inclined gauge A sloping gauge or plank graduated to record vertical heights above a permanent line or plane of reference. See STAFF GAUGE.

increment (groundwater) See GROUNDWATER INCREMENT.

index contour An important CONTOUR LINE which is marked more heavily than others and labelled at intervals along its length by a number which denotes its elevation above mean sea level. See STRUCTURE CONTOURS.

index map (or **location map**) A map or plan, showing the main surface features (towns, roads, rivers, etc.) of the area embraced in a geologic or hydrogeologic survey. The outline of any proposed works may also be shown. An index map may accompany a geological or engineering report. See SITE PLAN.

index, moisture See MOISTURE INDEX.

index, zero moisture See MOISTURE INDEX, ZERO.

indicator plants Plants that indicate, broadly, the underlying soil conditions, such as its qualities, deficiencies, salinity, and alkalinity. The plants may also indicate the approximate depth to zone of saturation and the climatic conditions of the area. See MARITIME PLANTS.

indirect flood damage All flood losses resulting from the interruption to services and production of goods, etc., as opposed to DIRECT FLOOD DAMAGE.

induced recharge of aquifer   Recharge of aquifer by inflow of stream water.

industrial water   Water used by industry for all purposes except that used by employees and public health. Industrial uses include steam raising, cooling plants and air conditioning. See POTABLE WATER.

infiltration   The slow movement of water through, or into, the pores or interstices of a soil or other mass; the absorption of water by the soil. It may be determined by an INFILTROMETER.

infiltration area (well)   The area or extent of water-bearing rocks penetrated in a well and which discharge water into it. See GRAVEL FILTER WELL.

infiltration capacity   The maximum INFILTRATION RATE of a soil or other porous material under specified conditions.

infiltration coefficient   The ratio of INFILTRATION to PRECIPITATION for a soil or substance under specified conditions.

infiltration ditch   See INFILTRATION TUNNEL.

infiltration diversion   A form of stream water diversion by the use of pervious conduits, or perforated pipes, which are set under the bed of the stream.

infiltration gallery   A construction to obtain a supply of water from a river where the banks consist of porous gravels or sands. A gallery or tunnel is driven close to and parallel with the river. The River Derwent supplies Derby (U.K.) with 13 500 m³/d (3 m.g.d.) by means of infiltration galleries driven in shales and gravels bordering the river. The galleries are 1·2 m (4 ft) in internal diam., lined with 230 mm (9 in) thick brickwork in which the vertical joints were left 6 mm (0·25 in) open. Their distance from the river varies from 3 to 15 m (10 to 50 ft). Variations of the method consist of pipes placed in a trench or small tunnel dug parallel to the river or in sinking wells in the porous banks. See INTAKE WORKS.

infiltration index   The INFILTRATION RATE, as determined from rainfall and runoff records. There are several different indices, depending upon the method of calculation.

infiltration rate   The rate at which water is absorbed, or seeps into or through the interstices of a soil or other porous material; may be stated in millimetres or inches per hour. It varies with the INFILTRATION CAPACITY of the soil and the rate at which water is applied. As an approximation for irrigation purposes, gravelly soils may take in water at the rate of 25 mm (1 in) in an hour, whereas a heavy clayey soil cannot absorb more than 8 mm ($\frac{1}{3}$ in) per hour.

infiltration tunnel (or **infiltration ditch**)   A tunnel, ditch, or gallery driven into the zone of saturation and along which water flows outwards by gravity into a well or water lodgment, or to the surface. See also INFILTRATION GALLERY.

infiltrometer   An instrument for measuring infiltration. The water during a test is applied artificially by either sprinkling or flooding. See PERCOLATION GAUGE.

inflow hydrograph (reservoir)   A graph showing inflow into a reservoir.

inflow-outflow method A SEEPAGE LOSS test in a water supply channel. A test section of the channel is taken and measurements made of flow into it and the corresponding flow out of the section. The seepage loss is the difference between the two flows with adjustments for rainfall and evaporation over the test period. During the test, channel levels are kept at the operating supply levels to ensure the normal seepage opportunity. See PONDAGE METHOD.

influence basin (of well) The basin-shaped depression in the water-table around a well due to withdrawal of water by pumping. See AREA OF INFLUENCE, CONE OF DEPRESSION.

influence of well, area of See AREA OF INFLUENCE (WELL).

influent seepage The seepage of water from a stream or channel outwards into the ground; movement OF GRAVITATIONAL WATER towards the zone of saturation. See EFFLUENT SEEPAGE.

influent stream A stream which continuously loses water by seepage to ground storage.

infra-red photography Photography on special films which are more sensitive to infra-red rays than light rays. Sometimes used in air surveys during misty weather. Can be used to locate waterlogged land and also the presence of freshwater intrusions in coastal waters.

Inglis's formula A British flood formula devised for India by Sir Claude Inglis; he considered that it covered almost all catastrophic floods recorded in that country:

$$Q = \frac{7000A}{\sqrt{(A+4)}}$$

where Q is the flood intensity in cusecs, and A is the catchment area in sq. miles. (Technical Report, 1943. Central Irrigation and Hydro-dynamic Research, Poona, India.) See FLOOD FORMULAS.

ingrown meanders The series of meander-like loops formed by a YOUTHFUL STREAM when it enlarges pre-existing bends.

initial abstraction The part of precipitation forming DEPRESSION STORAGE and INTERCEPTION.

initial detention The part of rainfall which does not appear as surface runoff or as infiltration during period of precipitation; includes evaporation, interception by vegetation, and depression storage.

initial rain Rainfall during the initial period of a storm before DEPRESSION STORAGE is completed. See RESIDUAL RAIN.

injection station The point or cross section at which an indicating solution is injected into a stream or channel to measure velocity and discharge. See DILUTION METHODS.

inlet (1) The upstream end of a conduit or any structure through which water flows. (2) A structure at the diversion end of a channel. (3) A surface connection to a closed drain. (4) A recess or bay in a lake, large river, or the shore of the sea.

innings An area of land reclaimed from a waterlogged or marshy lowland or from the sea. See LAND ACCRETION.

insequent drainage A drainage system in which young streams follow the irregularities of the ground, often on a nearly flat plain or upon superficial deposits of indefinite structure. See DRAINAGE PATTERN.

insequent stream A stream whose course is determined by factors of a minor nature which may be obscure; often applied to streams formed on the valley slopes of CONSEQUENT STREAMS and SUBSEQUENT STREAMS.

inspection chamber (or **manhole**) A pit from the surface down to a sewer, large water pipe, or siphon, which a man can enter for inspections and repairs. Deep manholes are often made of 0·9 m (3 ft) diameter precast concrete pipes fitted with step irons. Manholes may not be necessary for sewers larger than 0·9 m (3 ft) diam. through which a man can pass.

insulated stream A stream which neither receives water from the zone of saturation nor contributes water to it; a stream which is separated from the zone of saturation by an impermeable deposit. See PERCHED STREAM.

intake (of a well) The voids, cracks, or fissures in a water-bearing rock through which water passes into a well. See OPEN-END WELL.

intake area of aquifer The extent or area of outcropping permeable rocks from which an aquifer is fed with surface waters, such as streams and DEEP PERCOLATION. See RECHARGE OF AQUIFER.

intake heading See HEADWORKS.

intake of groundwater See GROUNDWATER INCREMENT.

intake tower (or **valve tower**) A concrete, stone, or cast-iron tower forming part of the outlet works of a storage dam. All water released from storage passes through the intake tower, which also houses the outlet control gates and valves of pipes drawing off water at different levels. See DRY VALVE SHAFT.

intake works The structures, channels, pipelines, etc. used to obtain a supply of water from a river, lake, or other source. Intake works are located where the banks, ground, and flow conditions are reasonably stable, with the minimum risk of erosion or silting. See SUBMERGED INTAKE, EXPOSED INTAKE.

integrated flow curve A flow graph showing accumulated volume of runoff at various discharges.

integrating meter An instrument for recording the total quantity of water which flows past it. See FLOW METER.

integration method (velocity) A mean velocity measurement of a stream at a vertical depth by means of a current meter. The meter is slowly lowered from surface to bed of stream and back to surface and the total number of revolutions of the wheel and time are recorded. See CURRENT METER.

intensity curve (rainfall) A curve showing relation of RATE OF RAINFALL to its duration.

131

intensity frequency of rainfall   The RATE OF RAINFALL which, on the average, is exceeded or equalled once in a given period of years. Also called RAINFALL RECURRENCE INTERVAL or RECURRENCE INTERVAL (RAINFALL).

intensity of rainfall   See RATE OF RAINFALL.

interception   The process by which precipitation is caught or retained by foliage and vegetation before reaching the ground.

interception channel (or **interception ditch**)   A ditch or channel excavated near roads or at the foot of slopes or at the top of earth cuts, to intercept and collect surface water; a CATCHWATER DRAIN. See GRIP.

interceptometer   A device to determine rainfall lost by interception; it is placed under trees or foliage and its catch of water compared with that of a rain-gauge placed on open ground without vegetal cover.

interference of wells   See WELL INTERFERENCE.

interflow   See SUBSURFACE RUNOFF.

interfluve   A ridge of land between drainage valleys or between streams. See DIVIDE.

intermediate belt   That part of the ZONE OF AERATION which lies between the CAPILLARY FRINGE and the BELT OF SOIL WATER; the contained water is sometimes termed INTERMEDIATE VADOSE WATER.

intermediate vadose water   The water contained in the INTERMEDIATE BELT.

intermittent filtration (or **land treatment**)   A method of sewage disposal in which the sewage effluent is spread over land and the use of field drains for removal of the water. See BROAD IRRIGATION.

intermittent spring   A spring from which the flow of water is variable and the discharge ceases during certain periods of the year.

intermittent stream   A stream which ceases to flow during certain periods of the year; it is usually perched above the water-table and flow can usually be correlated to the seasons or periods of rainfall.

internal drainage   A basin or area where runoff has no direct outlet to the sea. The Corangamite basin in western Victoria, Australia, is an example. See also CLOSED DRAINAGE.

internal drainge board   In Britain, a local drainage authority instituted by a river authority to improve the drainage of low-lying areas under its control. See WATER RESOURCES ACT 1963.

internal water   Sometimes applied to any water contained in the lower regions of the earth and below the zone of saturation.

interrupted protection (river banks)   Applied in river engineering to any form of intermittent bank protection. Alternate lengths of river bank are protected, leaving the intervening portions only lightly protected or even unprotected. The principle is similar to a series of groynes. It has been used successfully in New Zealand and in Victoria by the State Rivers and Water Supply Commission. Broadly, the objective is to secure the maximum possible bank protection for a specific outlay. See SADDLEBAG GROYNES.

interrupted stream   A stream which does not flow regularly throughout its course; a stream with perennial stretches with intervening intermittent or ephemeral stretches. See EPHEMERAL STREAM, INTERMITTENT STREAM.

interrupted water-table   A water-table with a marked difference in level in the vicinity of a fault or dyke or other obstruction to lateral flow of water. See GROUNDWATER DAM.

interstices   See VOIDS.

interstitial ice (or **subsurface ice**)   Ice occurring below the surface.

interstitial water   The water contained within the voids or interstices of a rock or soil. See CONNATE WATER.

intertropic front (or **intertropic convergence zone**)   A zone of heavy rains, local thunderstorms, and strong squalls; developed when an outbreak of cold air penetrates the DOLDRUMS and this calm belt is replaced by a narrow zone of violent interaction with equatorial air; location of severe tropical weather.

intrapermafrost   Layers of water within a mass of PERMAFROST.

inundation canal   A canal which depends for its supply upon the water level in a stream or river. It may or may not have some form of head regulator.

invert   An inverted arch, as at bottom of sewer, culvert, drain, tunnel, or channel (Figure S.2). See SOFFIT.

inverted capacity   The maximum rate at which an INVERTED WELL can remove surface or near-surface water by discharge through openings into deposits at its lower end. See DRAINAGE WELL.

inverted siphon (or **sag pipe**)   An irrigation structure which carries water discharges under railways, rivers, roads, or ground depressions. The inlet and outlet structures are usually of reinforced concrete, with a connecting tube or tubes of similar material. See SIPHON.

inverted well   A well in which the water flow is downwards; the water enters at or near the top and is discharged into a permeable deposit through openings at lower levels. See also RECHARGE WELL.

invert level   The level of the lowest part of an INVERT; the level which defines the elevation and slope of a channel, sewer, or drain.

Irish bridge   An open stone drain or ford carrying water across a road; a WATERSPLASH.

iron-clad catchment   See ARTIFICIAL CATCHMENT.

irrigable land   Arable land sufficiently low for irrigation. It may include areas and constructions not previously irrigated. See NON-IRRIGABLE LAND.

irrigating head   The flow of water required or available for irrigation; the flow rotated among a group of irrigators; the flow of water in a single farm lateral or distributed at a single irrigation.

irrigation   Spraying or causing water to flow over arable land for farming and to benefit crops. The work may involve excavating canals, laterals, and ditches, and constructing small dams and associated works. The rate of irrigation should not be excessive, or lime and nutrients in the soil may be lost by leaching. See WATER HARVESTING.

irrigation district An irrigation area, with definite geographic boundaries, under the control of a local authority, state department, or other body. See GROUP ENTERPRISE.

irrigation efficiency (or **distribution efficiency**) Usually expressed as the percentage of the total inflow into the channel system that is recorded as deliveries to the farms. The quantity of water that is actually delivered should be carefully measured to obtain reliable results. See EFFICIENCY OF FARM IRRIGATION, WASTE.

irrigation intensity That proportion of any given area which is irrigated.

irrigation lateral A ditch or channel branching off the main canal and carrying water to the farm ditches.

irrigation layout plan A scale drawing showing the entire arrangement for the irrigation of a farm; it may show quantity of water used or required, ground slopes, and elevations at key points, etc.

irrigation requirement (or **duty of water**) The amount of water required, including wastes (but excluding rainfall), for farming and production of crops; usually expressed as rate of flow per unit area, or as depth of water in a given period of time.

irrigation structure Any structure or appliance required for the regulation, conveyance, and measurement of water within an IRRIGATION DISTRICT.

isobars Lines on a map or weather chart connecting points or places of equal barometric pressure at a standard level.

isobath (water-table) A line on a plan connecting points at the same height or elevation above an aquifer or water-table. See WATER-TABLE CONTOUR PLAN.

isochion (or **isonival**) A line on a map or ground connecting points of equal water content of snow or of equal snow depths.

isochrone A chart or map of sewer system or river drainage showing a series of time lines; from origin to outlet of system, the transit time of water is indicated.

isohyetal map A map showing rainfall depth contours.

isohyetals (or **isohyets**) Lines of equal rainfall. The lines are marked at intervals by interpolation on maps showing a large number of stations at which rainfall observations are made. In the U.K. maps of this nature are issued annually in *British Rainfall* (q.v.), a Meteorological Office publication. In water engineering schemes, local records of rainfall are important; commonly one gauge per 200 ha (500 acres) of gathering ground is used. See BRADFORD GAUGE.

isomeric values These rainfall values are ISOPERCENTALS in which the monthly average is given as a percentage of the annual average.

isonival See ISOCHION.

isopachyte A line on a geological map connecting points of equal thickness in a specific rock stratum, aquifer, or other deposit; a contour of equal thickness. The depths or thicknesses for constructing isopachytes are obtained from boreholes, wells, outcrops, and perhaps geophysical surveys.

isopercental A line on a map connecting points of equal percentage of rainfall; each rain-gauge station shows annual or monthly rainfall as a percentage of annual average values for that station over a long period.

isopiestic line An imaginary line connecting points which possess the same static level; a contour of the PIEZOMETRIC SURFACE of an aquifer. See PIESTIC INTERVAL.

isopluvial line A line connecting all localities with the same PLUVIAL INDEX over a given period.

isotherm A line on a map or weather chart connecting points or places of the same mean annual temperature.

J

jet See ORIFICE.

jetting An hydraulic method of inserting WELLPOINTS or piles into sandy material, and used where a pile hammer might damage structures in the vicinity.

jetty (1) A ridge, dike, or construction of rock, piles, or other materials that jets or projects into a stream or into the sea at the mouth of a river for bank building or protection or to induce scouring. See GROYNES. (2) A landing stage or deck carried on piles from the water's edge.

joint-sealing material Mastic or bituminous material used for filling expansion joints in concrete flumes, siphons, and other hydraulic columns and structures. A pressure gun may be used when applying the material.

Joosten process A GROUTING method in which chemical fluids are injected into the ground through pipes in boreholes. The chemical grout strengthens and seals the soil, sands, or gravel and prevents the seepage of water, air, or gases into or from tunnels, shafts or other works. Sodium silicate is used to react, in the voids of the rock, with calcium chloride to form calcium silicate which acts as the sealing agent. See CHEMICAL CONSOLIDATION.

juvenile water (or **magmatic water** or **plutonic water**) Water derived from magmas or molten masses of igneous rock, during their crystallisation, or from lava flows in the form of steam.

K

kames Ridges of gravel deposited from the ice along the margins of a glacier. They lie parallel to the ice limits and indicate pauses in the retreat of the ice. Kames frequently enclose small lakes and some subsequently become peat bogs. See ESKERS.

Kaplan turbine A propeller type water turbine with blades of a pitch which can be automatically adjusted with the load to improve its performance. See FRANCIS TURBINE.

karst topography A region where limestone underlies the soil and the land is characterised by many SINKS, with irregular divides between them, disappearing streams, and solution valleys. See SWALLOW HOLES, TUBULAR SPRING.

karst water Sometimes applied to water with a high capacity for dissolving carbonates, as when seeping or flowing through limestone rocks. See CHEMICAL WEATHERING, SINKS.

keel blocks The central row of DOCKING BLOCKS in a dry dock.

Kelvin tube An appliance for determining the depth of water, consisting of a lead tube containing a chemical, placed inside a glass tube. The tube is dropped into the water and taken out immediately it touches the bed. The depth of water, given on a scale, determines the colour of the chemical changes.

Kennedy's critical velocity See CRITICAL VELOCITY.

kerb inlet A kerb with openings to convey storm water to a drain or gulley.

Keuper The Keuper of the Triassic in the U.K. consists of marls and sandstones. The coarse Keuper Sandstones are highly permeable and the best water, for quality and quantity, is obtained from wells under a thin cover of Keuper Marl or on the outcrop belt. The water is used by many industries especially around Birmingham. See CARBONIFEROUS LIMESTONE.

kid A FAGGOT or FASCINE.

kidding See FAGGOTTING.

kieselguhr (or diatomaceous earth) Sometimes found as an unconsolidated deposit in ponds and lakes and consisting of the skeletons of siliceous organisms, such as diatoms. A very porous and absorbent substance and often used in waterworks as a filtering material. See DIATOMITE FILTER.

kinetic head See VELOCITY HEAD.

knick point The head of a youthful valley carved by a rejuvenated stream; the point where the old and new stream profiles intersect; any sudden drop in the profile of a stream.

Kutter's formula An empirical formula expressing the value of the coefficient C in the Chézy formula, in terms of HYDRAULIC MEAN DEPTH R, FRICTION SLOPE S, and ROUGHNESS COEFFICIENT N.

Metric units:

$$C = \frac{23 + 1/N + 0.001\ 55/S}{1 + (N(23 + 0.001\ 55/S)/\sqrt{R})}$$

Imperial units:

$$C = \frac{41.65 + 0.002\ 81/S + 1.811/N}{1 + (N(41.65 + 0.002\ 81/S)/\sqrt{R})}$$

See MANNING'S FORMULA.

L

lacustrine   Pertaining to lakes.

ladder   The bucket ladder of a machine for underwater excavation. See DREDGER, ELEVATOR DREDGER.

lag (time)   See TIME LAG.

lagoon   (1) A shallow stretch of salt water parted from the sea by a ridge or barrier of sand. See ATOLL, BARRIER BEACH, COASTAL LAGOON. (2) A shallow natural storage connected to a river or sea. (3) A pond used for sewage treatment.

lagtime (storm)   The time interval between the beginning of rainfall excess and the maximum rate of runoff during a particular storm. See DESIGN FLOOD, PRECIPITATION, EXCESSIVE, STORMWATER, TIME LAG.

lake   A large inland body of water. It may or may not have a single direction of flow, and the water may be either fresh or brackish. Natural and artificial lakes often provide water for domestic and industrial use. See CRUSTAL-MOVEMENT LAKES, EVAPORATING BASINS, LANDSLIDE LAKES.

lake-floor plain   The exposed floor of a lake from which the water has drained or evaporated away. The deposits forming the floor are usually evenly stratified and fine-grained. See PLAYAS.

laminar flow   See STREAMLINE FLOW.

laminar velocity   The velocity of water in a particular channel below which STREAMLINE FLOW always exists, and above which the flow may be either streamline or turbulent.

lamphole   A small pit put down over the centre of a sewer to enable a lamp to be lowered on a string. A man looking along the sewer towards the lamp from a manhole can detect any damage to sewer or obstruction to flow. See INSPECTION CHAMBER.

land   Any portion of the solid surface of the earth's crust. It includes soils, rocks, watercourses, climatic conditions, and any earthworks and man-made improvements. See HYDROSPHERE.

land accretion   The reclamation of land from swamps, low-lying boggy areas, or the sea. The methods employed include drainage and pumping, elevation of ground by groynes, dumping of soil or mud, and the planting of reeds or maritime plants to induce the deposition of silt. Large scale reclamation schemes are usually carried out by local authorities or state departments. See BEDFORD LEVEL, HUTT RIVER RECLAMATION, INNINGS.

land drain   See AGRICULTURAL DRAIN.

land drainage   See DRAINAGE, INTERNAL DRAINAGE BOARD.

land drainage (pumping)   See PUMPING (LAND DRAINAGE).

land management   The management and planning of all measures for the protection and prudent use of land and soil for the maximum long-term benefit to agriculture, industry and the national economy. See AFFORESTATION, LAND PREPARATION, SOIL CONSERVATION SURVEY, WATER REQUIREMENT.

137

land preparation   Making the ground surface capable of more intensive and uniform application of irrigation water by levelling and reshaping of irregularities and slopes. See SOIL CONSERVATION.

land reclamation   See LAND ACCRETION.

landslide (or slip or slide)   The downward movement of rock and soil on embankments or hillside slopes; it may be slow or sudden. Movement may start because of an increase in water content, deep cuttings, and heavy blasting along the base. The usual remedies are drainage of slopes, lowering of water-table by pumping, and planting of trees or shrubs. Wet patches indicate accumulations of water and trenches are often cut across such areas and the water removed. A site with steep slopes may become a major problem when a reservoir has been filled, owing to changes in groundwater conditions and levels. The construction of the PANAMA CANAL was greatly hindered and costs increased by many slides. The Cucaracha slide involved over 4 million m³ (5 million yd³) of material. The high rainfall of the region was an important factor. See DAM, FAULT.

landslide lake   A lake formed along the drainage line in a valley which has become blocked by debris from a landslide. See ACCIDENTAL LAKE, GLACIAL LAKE.

land spring   A spring, the water from which is derived from a surface stratum. The water is affected by local variations of rainfall and is liable to be polluted. See SPRING.

land treatment   See INTERMITTENT FILTRATION.

Lapworth's chart   See DEACON DIAGRAM.

large-area floods   The flooding of a relatively large area by low intensity storms with a duration ranging from a few days to several weeks. See SMALL-AREA FLOODS.

large bore   See OPEN PIT.

large dam   Defined in the International Commission on Large Dams (I.C.O.L.D.) World Register of Dams as a dam which is either greater than 15 m (50 ft) high measured from the lowest portion of the general foundation area to the crest, or between 10 m and 15 m (33 and 50 ft), provided also that it has a crest length of not less than 500 m (1600 ft), or a reservoir capacity of not less than 100 000 m³ (80 acre feet) or the provision for a maximum flood discharge of not less than 2000 cumecs (70 000 cusecs). There are over 10 000 large dams in the world and the number is increasing rapidly. See RESERVOIRS (SAFETY PROVISIONS) ACT 1930; INTERNATIONAL COMMISSION ON LARGE DAMS (Appendix).

late maturity (river)   A stage of the erosion cycle characterised by: (1) major rivers with a MEANDER BELT roughly equal in width to the VALLEY FLOODPLAIN; (2) nearly all land at about the level of river grades; (3) rounded hilltops of low relief and poorly drained, gently rolling areas. See EARLY YOUTH.

lateral   A small irrigation channel branching off a main supply channel. See DIVISION GATE.

lateral abrasion (stream)   The cutting into and wearing away of the banks of a stream; especially active along the outer banks at curves. Along the inner banks water flow is sluggish and deposition of silt often occurs.

lateral canal   A canal excavated more or less parallel to a river which is unsuitable for navigation owing to high velocity or another reason. See CANALISATION.

lateral-flow spillway (or **side-channel spillway**)   A SPILLWAY used on dam sites located in narrow valleys or canyons; the water passes over the spillway crest into a channel and then flows out parallel to the crest, that is, the initial and final flow are approximately at right angles to each other.

lateral moraines   See MORAINES.

lateral pipes (irrigation)   Portable pipes which convey the water from a HYDRANT or STAND-PIPE on the main to the positions where the water distributors are located. The remote end of each pipe is closed by a 'stop-end' or plug. A shut-off valve is often used to enable part of the system to be moved while irrigation continues along other sections. See MAIN PIPE, RING MAIN.

lateral storage   See BANK STORAGE.

lateral stream   A stream which flows in the depression along the edge of a lava flow in a valley. See TWIN LATERALS.

late youth   A stage of the erosion cycle often characterised by (1) efficient and well-integrated drainage of the area covered by the stream system; (2) alternating spurs in most stream valleys and (3) pronounced relief and sharp interstream divides. See LATE MATURITY.

lavants   Streams formed by VALLEY SPRINGS. See BOURNES.

law of continuity (flow of water)   States that the inflow of water to a given reach of a river or storage during a given period is equal to the outflow during the same period plus or minus any changes in storage volume.

lay-flat tube irrigation   Irrigation by means of jets of water issuing from small holes punched in thin-walled lay-flat polythene tubes. The tubes are laid down along the centre of beds, one end connected to the mains and the other end closed. The throw of the jets can be varied by a standpipe tap which adjusts the head of water. Water is applied at a high rate. See LOW-LEVEL SPRINKLER IRRIGATION.

leaching   The flushing out of excess salts from irrigation land, usually by applying large amounts of water over the affected area for extended periods. See SALTING.

leaching requirement   The proportion of the applied water that is used to leach the salt out of the soil.

lead line   See SOUNDING LINE.

leat (or **mill stream**)   A stream, or any open water course, used for power purposes in a mill, mine, farm, etc.

ledges   Platforms formed along a coast, composed of soft and hard beds, by the attack of storm waves; the ledges are formed on the hard rocks by the accelerated erosion of the overlying soft beds.

leech See LIMPET.

left bank The left hand bank of a river facing downstream (Figure A.4).

length (stream) The linear dimension measured along the course of a stream; usually between two specified points or cross sections. See REACH, STREAM PROFILE.

length-of-run (irrigation) The length along which irrigation water is conveyed in furrows or by flooding from the head channel.

levee (or **levee banks**) (1) An embankment to protect land from inundation during river floods or to control or confine river flow. The core may or may not be impervious. An artificial levee may be made wide enough at the crest to carry vehicles or to provide access in flood time; access ramps and turning places may also be provided. (2) A NATURAL LEVEE or river bank formed by the deposition of silt during flood periods. It forms elevated ground bordering the floodplain of a river. Natural levees are often increased in height by artificial means to protect the plain from overflow during river floods. See EMBANKMENT, RING LEVEE.

levee, natural See LEVEE.

level course See STRIKE.

level recorder An automatic device which records the water level in a stream or canal; operated by pressure or by a float. See GAUGE.

level terrace An embankment or ridge of earth formed along the absolute contour of the ground. Generally used where outlet channels are not practicable or on permeable soils where water conservation is very important. See GRADED TERRACE.

Lias This formation is divided into the Lower, Middle, and Upper Lias, and the outcrop extends across England from the Yorkshire to the Dorset coast. The beds yield useful supplies of water which is generally soft in the Upper Lias and hard in the Middle and Lower Lias. See GREAT OOLITE.

life buoy A device to keep a person afloat when he is in danger of drowning. See NUN BUOY.

life of reservoir The period in years that a reservoir yields useful supplies of water. Mainly, it varies with the amount of silt deposited by streams, etc., and the remedies applied. Some storages in India, China, and U.S.A. will only last about 50 years. It is estimated that the life of the Eildon reservoir in Australia (capacity 3392 million m^3 or 2 750 000 acre feet) is about 17 000 years at the present rate of siltation. See SILTING.

lift (1) The vertical distance through which the water must be raised from a river or channel. (2) A power-operated hoist which elevates or lowers vessels from one reach to the next where a lock is not available. (3) The vertical distance a vessel is elevated or lowered when passing through a lock.

lift gate A LOCK GATE which opens by moving vertically upwards. See PENNING GATE.

lift pump An earlier type of pump consisting of a column of rods operating a piston within a cylinder or working barrel. A reciprocating motion

140

causes the water to be raised to the surface, or a higher level, by a series of lifts. The piston on its down-stroke (with the flap valves open) admits water to the upper section of pipe. On the up-stroke the valves close and the water is lifted in the pipe column. See CENTRIFUGAL PUMP, FORCE PUMP.

lighthouse A tower or structure supporting and protecting a recognisable light signal at the top which is transmitted to serve as a guide and warning to ships at night; erected at some important point on a coast or at the entrance of a port. The first modern structure was the Eddystone lighthouse designed and constructed in 1757–59 by John Smeaton, the first man to call himself a civil engineer.

lightship A ship moored at a site near the coast where it is not possible to build a LIGHTHOUSE. It is fitted with a powerful light and foghorn to warn or guide ships at sea. Modern lightships also have radio direction-finding apparatus and radar. The ships are manned or unmanned, the latter being fitted with automatic lanterns. The lanterns are normally visible for a distance of at least 16·1 km (10 miles). The first English lightship (1732) was moored on the Nore sandbank in the Thames estuary. It was replaced in 1844 and again in 1936.

limestone sinks See SINKS.

limits of oscillation The valley area or width within which a river has ranged within historic times. See MEANDER BELT.

limnology Branch of hydrology pertaining to ponds, lakes, and inland waters; the study of life in such waters.

limpet (or leech or limpet dam) A small OPEN CAISSON designed to fit against a dock wall which requires repairs. It is lowered into the water by crane and then pumped out. The open top provides access to the cais-

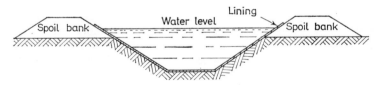

Figure L.1. Lined ditches

son. Repair work with a limpet is much quicker and more convenient than in a diving bell or diving suit.

limpet dam See LIMPET.

lined ditches (or cement ditches) Ditches with linings usually 50 mm (2 in) thick and made of a sand–cement mixture in the ratio of 3:1 to 5:1. They are usually trapezoidal in cross section, with side slope in the ratio of 1:1 or $1\frac{1}{2}$:1, and the depth of water is about equal to the width of the bed (Figure L.1).

line of breakers (waves)   See BREAKERS.

line of creep (or **path of percolation**)   The course or direction taken by water along the contact surface between the foundation soil and the base of a dam or other structure. See PIPING, TUNNELLING.

lining   A layer or cover over the whole or part of the perimeter of a conduit or a reservoir to minimise seepage losses, resist erosion, withstand pressure, and in general improve flow conditions. On canals, a lining of clay, concrete, polythene, or other material may be used. See REVETMENT, RIPRAP.

lip   (1) A small structure, such as a wall, to break the flow on the downstream end of the APRON. (2) Part of a dam.

liquefaction   Applied to a substance, such as saturated sand, with a VOIDS RATIO higher than the critical voids ratio, which tends to liquefaction when subjected to sudden loading or vibration. See QUICK CONDITION, QUICK-SAND.

liquid level recorder   See WATER STAGE RECORDER.

liquid surface profile   See SURFACE PROFILE.

listening methods   Methods used to detect underground leakage of pipes. Devices used include listening rods and stethoscopes fitted with earphones. See WASTE.

lister plough   An earth-moving machine in which the blades throw the soil in opposite directions, thus forming a series of alternate ridges and furrows. See BASIN LISTER.

lithological map   A GEOLOGICAL MAP showing the main rock types of an area, their general distribution, and the principal lines of faulting and folding. Where local water supplies are involved, the arrangement of the pervious and impervious rocks and any fault lines are important. The rock types may be denoted by signs or symbols or by colour. Water-bearing rocks are sometimes shown by a light wash of blue. See WATER PROSPECT MAP.

littoral current   A flow of water parallel to the seashore. See SHORE CURRENT.

littoral drift   The transport of beach material along a shore belt by marine current and waves. Interference with the normal littoral drift by engineering works may lead to difficulties, such as heavy erosion. See LONG-SHORE DRIFT.

livestock reservoir   See TANK, EARTH, FARM RESERVOIR.

Lloyd Davies formula   Sometimes used in the U.K. to calculate the runoff; the figure may be used to determine the size of sewers. Given as: runoff water in $ft^3 = 60 \cdot 5 \times$ area drained in acres $\times$ rainfall in in/h $\times$ impermeability factor.

load (stream)   The weight or volume of material carried or transported by a stream per unit of time. The load may be moved (1) in SUSPENSION when consisting of mud, silt, or sand; (2) in chemical or colloidal solution, when consisting mainly of calcium and magnesium carbonates and small quantities of other salts, such as sodium chloride; (3) by TRACTION when pebbles and boulders move along the bed of the stream, and (4) by

142

SALTATION or the intermittent movement of heavy particles by turbulence and eddy currents. See TRANSPORT COMPETENCY.

loaded filter (or **reversed filter** or **weighed filter**) A GRADED FILTER placed at the foot of an earth dam or similar hydraulic structure. Its weight stabilises the toe of the dam and its permeability prevents any water being trapped under pressure.

loader, front-end See FRONT-END LOADER.

local base level The level of a lake or body of water into which a stream flows. See TEMPORARY BASE LEVEL.

local inflow A flow of water into a stream between two gauging stations.

location map See INDEX MAP.

lochs A lake; a bay or arm of the sea (Scotland).

lock A chamber between two stretches of a canal, with an upper and lower gate, through which small ships or barges can pass in either direction. See DOUBLE LOCK, HEAD BAY.

lockage The water which flows from the upper to the lower reach of a canal during the passage of a vessel through the lock.

lock bay See CHAMBER.

lock cut (or **cut**) A short length of canal excavated near a river to enable boats to by-pass a weir and pass through a lock.

lock gate A gate or movable barrier which separates the water in an upstream or downstream stretch from that in the lock chamber.

lock paddle A SLUICE to enable a lock chamber to be filled or emptied. See PADDLE.

lock sill (or **clap sill** or **mitre sill**) The step in the floor of a lock CHAMBER forming a stop against which the gates bear when closed.

lodge A pumping station or the storage sump near the pit-bottom of a mine or other pumping centre. See SUMP.

log A float and line for measuring the speed of a ship; a record of the rate of a ship's speed. See INTEGRATION METHOD.

log chute (or **log-way**) A by-pass maintained around or through a dam for the passage of driftwood or logs.

log line (or **sounding wire**) A length of flexible wire with a weight at one end for measuring the depth of water in streams, canals, etc. See HAND LEAD, SOUNDING LINE.

logs, bush See BUSH LOGS.

log training wall A TRAINING WALL to prevent erosion of river banks, consisting of pyramidal trestles built of logs about 150 mm (6 in) thick and about 3·7 to 6 m (12 to 20 ft) apart, with other logs wired from trestle to trestle on the river side. The wall induces the deposition of shingle, especially during floods, and an embankment is formed. See WOOLF SYSTEM.

log-way See LOG CHUTE.

longitudinal crack A form of crack in an earth dam or embankment which develops more or less parallel to the axis of dam. Not as dangerous as a transverse crack but can be troublesome. See SETTLEMENT.

longitudinal section (or **longitudinal profile**) A section drawn vertically through the centreline of a canal, conduit, dam or pipeline. It shows levels along the original and final ground. See PROFILE.

long-shore drift The transport of material by waves running obliquely up the beach. The dominant direction of the drift is determined by the prevailing winds, of moderate intensity, blowing obliquely on the beach. See LITTORAL DRIFT.

loose rock dam A rubble dam; a dam constructed with rock material without the use of mortar. See ROCK-FILL DAM.

loose stone Large or small stones which are not restrained by wiremesh, timber work, or vegetation. When used for groynes and river bank protection, they are usually secured to prevent dislodgment by water flow and waves. See STONEMESH CONSTRUCTION.

losses, flood See FLOOD LOSSES.

loss of head See FRICTION HEAD, HYDRAULIC FRICTION.

loss rate, flood See FLOOD LOSS RATE.

lost head See FRICTION HEAD, HYDRAULIC FRICTION.

lower critical velocity See HIGHER CRITICAL VELOCITY.

Lower Greensand The Lower Greensand of the Cretaceous in the U.K. consists of permeable sands and retains plentiful supplies of good soft water which is much used in Bedfordshire and Cambridgeshire. Some deep bores which penetrate the Lower Greensand, below the Chalk and gault, yield large quantities of artesian water. See MAGNESIAN LIMESTONE, UPPER GREENSAND.

lower high water (L.H.W.) The lower of two high waters of any lunar day.

lower low water (L.L.W.) The single low water, occurring daily during periods when the tide is diurnal, is considered a lower low water; lower of the two low waters of any lunar day.

lowest low water (L.L.W.) The lowest known water level.

low-heat cement A cement with a lower heat generation during setting than ordinary Portland cement. It is often used in concrete dams because the lower temperature rise reduces the risk of consequent shrinkage cracks and leakage. Low-heat cement (BS 1370) is a coarse-ground low-lime cement. See TRIEF PROCESS.

low level sprinkler irrigation Irrigation by means of fine jets of water issuing from small holes drilled in small-bore rigid plastic pipes laid at ground level. The method gives a larger volume of wetted soil and is less liable to blockage than TRICKLE IRRIGATION. Adjustment of the pressure head, by means of a hand valve, allows the trajectory of the jets to be altered. See LAY-FLAT TUBE IRRIGATION.

low water (L.W.) The lowest water level within the period specified; lowest level reached by a falling tide.

lysimeter An instrument to measure the percolation of water through soils and to determine the soluble constituents removed.

M

macropores Rock pores greater than 0·005 mm. See MICROPORES, VOIDS.

maelstrom A WHIRLPOOL on the west coast of Norway; any great whirlpool.

magmatic water Water released by molten igneous rocks or magma during the process of cooling and consolidation. See JUVENILE WATER.

Magnesian Limestone The Magnesian Limestone of the Permian in the U.K. consists largely of marine dolomitic limestones, marls, and gypsum. The rock is compact and of low permeability but in places the joints and fissures provide storage for water and many wells obtain their supplies along the fissured zones. The water is hard and often requires softening and other treatment. See KEUPER.

magnetic method A GEOPHYSICAL PROSPECTING method which detects changes or anomalies in the magnetic field. It may be used to trace geological structures or igneous intrusions or fissured zones forming aquifers. Igneous dykes which form GROUNDWATER DAMS may be located and traced due to the presence of magnetite in the rock.

main canal See CANAL, MAIN IRRIGATION.

main drain A land drain or sewer which leads directly to its point of discharge or OUTFALL.

main pipes (irrigation) The pipes which convey the water from the storage or pumping source to the distribution or discharge points. They may be carried on concrete or brick piers, laid on the surface, or buried about 0·6 m (2 ft) below the surface. Main pipes are usually aluminium, asbestos cement, plastics, iron, or steel, the last two often protected with bitumen layers. See RING MAIN.

mainspring A spring the water from which is derived from an underlying stratum and is thus unaffected by local variations of rainfall. The water is usually potable, but pollution may occur at the outlet.

maintenance The continuous inspection and repair of water supply schemes and river control works.

maintenance of groynes The regular inspection and maintenance of groynes, particularly during the early stages until they are well stabilised. The head or water end is particularly vulnerable to undermining and other damage, and repairs are carried out as soon as possible. See AUXILIARY PROTECTION (GROYNES).

maintenance works (land drainage) Works and activities to restore the ravages of nature and to maintain the efficiency of a river system for optimum drainage and plant growth. They may include (1) deepening of channels by removal of silt and shoals, (2) reduction of silting by removal of obstructions, such as fallen trees and weeds, and (3) prevention of damage to banks and bed by fencing and protective works. See GROYNES, LEVEE.

main water-table (or **phreatic surface**) The surface or water-table of the ZONE OF SATURATION; not applicable to PERCHED WATER-TABLE.

major dam See LARGE DAM.

major divide See DIVIDE.

management See CATCHMENT MANAGEMENT, GRAZING MANAGEMENT (catchwaters), RIVER CONTROL, WATER CONSERVATION.

Manchester Ship Canal A navigation canal, constructed 1887–94, connecting the Mersey to Manchester (U.K.). It is 58 km (36 miles) long, 8·5 m (28 ft) deep, and 52 m (170 ft) wide.

manhole A small pit or chamber which allows a man to enter and examine a drain, sewer, or large water supply pipes and siphons. See INSPECTION CHAMBER.

manhole cover A movable cast-iron or steel plate over a manhole or inspection chamber. The plate fits into a cast-iron frame bedded in concrete. In the case of foul drains, the cover is usually formed with a seal to prevent the escape of foul gases.

manned cableway (gauging) A cableway where the operator controls the CURRENT METER from a car which travels on the cable. A special grip enables the car to be moved along the cable by hand.

Manning's formula A formula expressing the relation between average velocity of water flowing in an open channel V, HYDRAULIC MEAN DEPTH R, and slope of the HYDRAULIC GRADIENT S, containing an empirical constant N, as used in KUTTER'S FORMULA and varies according to CHANNEL ROUGHNESS:

Metric units

$$V = \frac{1}{N} R^{0.67} S^{0.50}$$

Imperial units

$$V = \frac{1 \cdot 486}{N} R^{0.67} S^{0.50}$$

See CHÉZY FORMULA, KUTTER'S FORMULA

manometer A 'U'-shaped tube for measuring pressure differences in a water pipe. The tube contains a liquid which does not mix with water such as kerosene, and its two ends are connected to the two points in the pipe between which the difference in pressure is required.

map See PLAN.

map, hydrogeological See HYDROGEOLOGICAL MAP.

map symbols The letters, numbers or other signs marked on a map to denote rock types or soils, or indicate structural features. The standardised symbols used on a particular map are printed in the margin with the definitions. See LITHOLOGICAL MAP.

marigraph A gauge which records the level of the tides at a tidal observation station.

marine borers Animals, such as molluscs and crustaceans usually inhabiting warm waters, which damage timber structures by eating into them; the shipworm and gribble are typical and in favourable water penetrate almost any wood. The usual remedy is treatment with creosote and similar substances, or the use of concrete as a structural material. See SHEATHING.

marine climate The climate prevailing over land areas near the sea. See CLIMATE, OCEANIC.

marine surveying (or **marine hydrographic surveying**) The study and mapping of the ocean floor, and the charting of current, etc. See HYDROGRAPHY, OCEANOGRAPHY.

maritime plants Plants which grow in salty conditions such as foreshores. They often serve a useful purpose in reducing or preventing scour, and may be cultivated to form a protective covering to soil or rocky surfaces. See REVETMENT.

marsh A flat, usually low-lying area of land, often covered with water and supporting a native growth of rushes, reeds, or grasses. The waterlogged condition may indicate that the water-table is at or slightly above ground level. See SWAMP, WATERLOGGED.

masonry dam A dam constructed with stone bedded in mortar. The 43·9 m (144 ft) high Vyrnwy Reservoir in North Wales (U.K.) is one example of this kind of dam.

mass centre In hydrology, usually refers to the centre of gravity of the MASS CURVE.

mass curve See MASS DIAGRAM.

mass diagram (or **mass curve**) A graph showing cumulative flow quantities, such as the integration of a time–flow curve. Each point on the graph is the sum of all preceding flows. The diagram is of value in water storage studies. See DISCHARGE MASS CURVE, PRECIPITATION MASS CURVE, RESIDUAL MASS CURVE.

mass discharge curve See DISCHARGE MASS CURVE, MASS DIAGRAM.

massive buttress dam See BUTTRESS DAM.

mass movement See LANDSLIDE, EARTHQUAKE.

materials surveys The investigation and appraisal of construction materials available at a proposed site. For example, a large dam would require vast quantities of materials, the winning and placing of which would represent a high proportion of the total cost. The surveys would include the availability, at a reasonable cost, of rock for use as rockfill and as coarse aggregate for concrete, sands for concrete, fine aggregate, and also clays and shales suitable for earthfill. The dam design is often adapted to make the maximum use of local material, and this is only possible after careful materials surveys. See CONSTRUCTION MATERIALS, DAM SITE INVESTIGATIONS.

mattress A layer or blanket of brushwood, interwoven or otherwise lashed together, and weighed down with concrete blocks or stones to reduce erosion on the bed and banks of rivers or used with groynes. See FASCINES, REVETMENT.

mature river A river with a velocity just sufficient to transport sediment delivered by the tributaries; a river with a meandering course, absence of lakes and waterfalls, and which has attained a PROFILE OF EQUILIBRIUM. See OLD RIVER.

maximum capacity (of well) The maximum rate at which water can be withdrawn from a well and expressed in cumecs or cusecs, litres per minute or gallons per minute. See ECONOMIC YIELD, TOTAL CAPACITY (OF WELL).

maximum envelope curve See ENVELOPE CURVE.

maximum possible flood The maximum flood that theoretically can occur at a given area, assuming simultaneous occurrence of all possible contrary factors, and during present climatic and geologic conditions.

maximum possible rainfall The rainfall of a certain amount and duration that theoretically could occur in a basin during the present climatic conditions.

maximum probable flood The heaviest flood that can reasonably be expected to occur along a river or selected stretch of a river. See FLOOD (return period).

maximum probable rainfall The rainfall of a certain amount and duration that can reasonably be expected to occur under existing climatic conditions, in a given basin or drainage area.

maximum rainfall intensity The rate of rainfall during any period when the rate is both maximum and uniform; the period is specified. See PRECIPITATION, EXCESSIVE, RATE OF RAINFALL.

meadow strip A strip of grassed land or sloping field, usually much larger than a GRASSED WATERWAY and often used as a terrace outlet channel. In addition to yielding a hay crop, the strip acts as a shallow, broad water channel for runoff.

mean depth The average depth of water in a stream or channel, equal to the cross-sectional area of water divided by its surface width. See HYDRAULIC RADIUS.

meander The loop or curve taken by a river as it flows over flattish areas; generally occurs when a river has cut down nearly to BASE-LEVEL.

meander belt The valley area enclosed by lines drawn tangentially to the outer sides of loops or bends of a river. The meanders of some large rivers have now been trained between embankments and walls along certain sections, to prevent further encroachment on adjoining land areas. See LATE MATURITY.

meander concertina Occasionally a MEANDER MIGRATION is arrested by a natural or artificial obstruction on the river (e.g. a bridge), which results in a crowding-up of the meanders upstream of the obstruction (Figure M.1).

meander, full See FULL MEANDER.

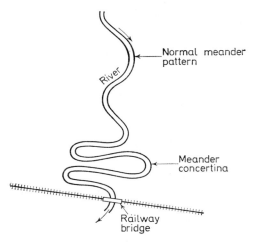

Normal meander
pattern

River

Meander
concertina

Railway
bridge

Figure M.1.   Meander concertina

meander length   The length between corresponding points at the outer sides of successive fully formed river meanders.

meander migration   The tendency of river meanders to move slowly downstream. A comparison of old and new maps of the same river stretch often shows this downstream shift of the meanders.

meander ratio   The ratio expressing relation of MEANDER WIDTH to MEANDER LENGTH.

meander width   The width or extent of swing of a fully formed meander measured from midstream to midstream. Also called AMPLITUDE.

mean higher high water (M.H.H.W.)   The mean level of higher high waters over a long period or within the period considered.

mean high water (M.H.W.)   Mean level of high waters over a long period or within the period considered. With semi-diurnal or mixed tides, only higher high water levels are included; in this case, mean higher high water is the same as mean high water.

mean lower low water (M.L.L.W.)   Mean level of lower low water within the period considered or over a long period.

mean low water (M.L.W.)   Mean level of low waters within the period considered or over a long period. With semi-diurnal or mixed tides, only lower low water levels are included; in this case, mean lower low water is the same as mean low water.

mean sea level   The level of the surface of the sea if it was stationary with no tides; the average sea level determined by measurements at equal intervals of time over a long period. See NEWLYN DATUM, ORDNANCE DATUM.

149

mean tidal range   The difference in level between mean low water and mean high water; the neap range and spring range are the near ranges of NEAP TIDES and SPRING TIDES, respectively.

mean tide level   The level midway between MEAN HIGH WATER and MEAN LOW WATER.

mean velocity   (1) Mean velocity for a REACH of a stream obtained by dividing discharge by average cross-sectional area of the reach. (2) Mean velocity at a given section of stream obtained by dividing the discharge by the wetted area at that section. See CURRENT METER.

mean velocity on a vertical   Obtained by slowly moving the current meter at a specified rate along the vertical, or by taking the mean of the velocity at two or more points on the vertical.

mean velocity point   The point lying between bed of channel and water surface, at which the velocity is equal to MEAN VELOCITY ON A VERTICAL.

mean water (M.W.)   The average level of the surface of still or flowing water. See MEAN SEA LEVEL.

measuring section   The section used for discharge measurements. See GAUGING STATION.

measuring weir   A structure or device for measuring rate of flow of water; consists of a triangular, rectangular, or trapezoidal notch in a thin vertical plate and normally placed at right angles to the flow. The rate of flow is calculated from the depth of water in the notch. See CIPOLLETTI WEIR, RECTANGULAR WEIR, TRIANGULAR NOTCH.

mechanical control   See ENGINEERING PRACTICES.

medial moraine   The middle moraine formed when two glaciers join; it consists of two LATERAL MORAINES.

meeting post (or mitre post)   The vertical timbers at the outer end of a pair of LOCK GATES, mitred so the gates fit tightly when closed. See LOCK SILL.

meiyu   Literally means plum rain in Chinese and refers to the season when heavy rains are required to ripen the plum. This season is about June and July and the term is applied in the Provinces south of the Yangtze. See BAI-U.

membrane   A fine lining placed on small farm dams or water supply channels to reduce seepage.

Menard pressure permeameter   An appliance which measures directly the permeability to water or gas of any rock in a borehole. The field determinations are confirmed by laboratory tests.

mere   A pool, pond, or lake. In areas where local subsidence occurs due to shallow mining and other causes, meres are common and are continually extending. This applies to parts of the salt mining area between Nantwich and Northwich (U.K.).

Mersey tidal current (or Mersey bore)   See TIDAL CURRENT.

meteoric water   The water which falls as rain or is derived from dew, hail, or snow. See PRECIPITATION.

meteorograph A self-recording instrument or apparatus for recording meteorological phenomena, such as temperature, pressure, and humidity; often applied to apparatus used for recording data in upper atmosphere, especially for weather forecasts.

meteorology The scientific study of the atmosphere and its phenomena; mainly pertaining to WEATHER and CLIMATE. See HYDROMETEOROLOGY.

meteorology, synoptic A general study or survey of atmospheric processes based on weather conditions prevailing over a widespread area at a given time.

meter A device placed in a pipeline or channel to measure the rate of flow of water. See CURRENT METER, DETHRIDGE METER, FLOW METER.

method of measurement. In U.S.A. and British earth dam contracts, the term CUT YARDS applies to measurement of volume as excavated. FILL YARDS is applied to materials as placed, compacted, or dumped in dam embankments. Also called COMPACTED YARDS. Note that a YARD in contractor's jargon means one cubic yard, and that volumetrically one cut yard, one fill yard, and one compacted yard are equal to 0·765 cubic metre.

microclimate The climatic conditions pertaining to a particular place and caused by local modifications in exposure and altitude.

microclimatology The science or study of MICROCLIMATES.

micrometeorology The study of METEOROLOGY pertaining to small areas, such as towns, cities, catchment areas, or a river system.

micropores Rock voids smaller than 0·005 mm. If the greater part of the total porosity of a rock consists of micropores the passage of water through it is inhibited. See MACROPORES.

micro-strainers A proprietary name for a drum strainer often used in the preliminary treatment of water containing microscopic organisms. The raw water passes inside the drum and outwards peripherally through a finely woven metallic fabric. The washing is continuous and the drum self-cleaning. Usually sited at a point before the water passes on to the SLOW SAND FILTER. See COARSE STRAINERS.

mid-bay bars Banks of sand, silt, or gravel formed along the middle of bays. See BAY-HEAD BARS.

middle third That portion of the thickness of a GRAVITY DAM which is one third of the total thickness and is located centrally along it. For the stability of a gravity dam one criterion is that the line of resistance must lie within the middle third, when the storage is both full and empty. See CENTRE OF PRESSURE.

middle thread A line along the course of a stream which would represent the final flowing of its waters immediately before cessation; a legal term.

migration, of meanders See MEANDER MIGRATION.

mildly brackish See BRACKISH.

millibar Unit of atmospheric pressure used in meteorology; equals 10^2 N/m^2, or approximately 1/32 in of mercury.

mills See MOULINS.

mill stream See LEAT.

mill tail See TAIL-RACE.

mineral spring A spring with a relatively high content of dissolved mineral matter, and often named after the dominant mineral.

miner's inch A rate of flow of water; originally the discharge from a 1 in square orifice under a specified head. The unit is used mainly in western U.S.A., where its volumetric flow value has been fixed by statute in various states. (One inch = 25·4 mm.)

minimum envelope curve See ENVELOPE CURVE.

minimum speed of response The minimum speed at which the angular motion of the rotor of a CURRENT METER remains uniform and continuous. See SPIN TEST.

minor dam Sometimes defined as a dam constructed for small country town water supply schemes; a dam of smaller size and capacity than a LARGE DAM. A DESIGN RECURRENCE INTERVAL of 500 years is sometimes recommended for minor dams where surcharge is likely to cause loss of life. See DESIGN FLOOD.

minor watershed (U.S.A.) A subdivision of a watershed which is convenient for investigation, description, and detailed planning of water and soil conservation measures; a subdivision of a drainage area of unspecified size. See WATERSHED.

'misfit' stream A stream which has lost its headwaters by RIVER CAPTURE and has diminished in size and is manifestly too small for the valley in which it now flows.

mist See FOG.

mitre post See MEETING POST.

mitre sill See LOCK SILL.

mixed-flow turbine An inward-flow reaction water turbine in which the water pressure acts both radially and axially on the runner vanes.

mixed tides Tides which change, according to the declination of the moon, from diurnal at some periods to semidiurnal at others.

model analysis The study of scale models of hydraulic structures, the measurement of the heads and discharges of water, and finally the interpretation of the results to predict the characteristics of the full-size or actual structures. Model analysis is used for dam spillways, river improvement works, harbours, and so on. See HARBOUR MODELS.

moderately brackish See BRACKISH.

modified Pacific type (coastline) Often applied to the coastline of Asia which lies behind a chain of islands along the western shores of the Pacific Ocean. See PACIFIC TYPE.

modified velocity The velocity after adjustment for ANGULARITY CORRECTION and DRIFT.

modular flow (weir) The flow through a flume or over a weir when the level upstream is independent of the level downstream. See DROWNED FLOW.

modular limit The SUBMERGENCE RATIO when the downstream level just begins to affect the flow.

module A contrivance for delivering water at a constant head through an opening 1 in (25·4 mm) high of variable width. See MINER'S INCH.

moisture Water diffused in small amount as vapour or liquid in rocks, soils, or atmosphere. See HUMIDITY.

moisture adjustment Modification of observed rainfall in a storm, over a drainage area under study, by use of ratio of estimated maximum precipitable water to actual precipitable water; may be used in estimating MAXIMUM PROBABLE RAINFALL. See DESIGN FLOOD.

moisture deficiency, field See FIELD MOISTURE DEFICIENCY.

moisture equivalent See CENTRIFUGE MOISTURE EQUIVALENT.

moisture equivalent, field See FIELD MOISTURE EQUIVALENT.

moisture, field See FIELD MOISTURE.

moisture gradient The rate of change in moisture content of a soil with depth.

moisture index (or **Thornthwate index**) One which indicates the adequacy of precipitation for the requirements of plants.

moisture index, zero The MOISTURE INDEX when rainfall, analysed monthly, is just sufficient to meet maximum transpiration and evaporation over one year.

moisture movement Applied to the increase in length or volume of a material with increased moisture content. See BULKING, SWELLING PRESSURE.

moisture regime (or **water regime** (irrigation)) A crop is said to be subject to a specific 'moisture regime' when the water content of the soil is permitted to fluctuate only between FIELD CAPACITY and a predetermined state of dryness. The regime is said to be 'dry' where the fluctuation is large, and 'wet' where the fluctuation is small. Maximum plant growth is obtained when the soil is maintained near to field capacity.

moisture status A term sometimes used for the wetness or dryness of a CATCHMENT AREA with particular regard to its capacity for absorbing rainfall and preventing this rainfall from becoming runoff.

mole See BREAKWATER.

mole drainage (or **moling**) The draining of a stiff clay by forming 75 to 150 mm (3 to 6 in) diam. tubular channels in the soil at a depth of from 0·6 to 0·9 m (2 to 3 ft) by means of a MOLE PLOUGH. The method is much quicker and cheaper than laying ordinary agricultural drains (Figure M.2).

mole-pipe drainage Drainage of an area by a combination of DRAIN PIPES and MOLE DRAINAGE. The pipes provide the main escape channels for the water; the mole drains provide secondary routes to the pipes and also fracture the soil to increase the rate of percolation. See SUBSOIL DRAINAGE.

mole plough (or **mole plow** (U.S.A.)) An excavating machine or plough consisting of a vertical knife blade with a horizontal bullet-shaped bottom member, from 75 to 150 mm (3 to 6 in) in diam., which is hauled through the soil for MOLE DRAINAGE. The plough is pulled uphill along the course

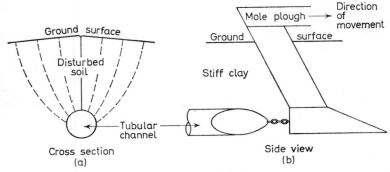

Figure M.2. Mole drainage

of the desired drains by a caterpillar tractor. Mole drains can also be used to lay copper or polythene water pipes. (Figure M.2).

mole plow See MOLE PLOUGH.

moling See MOLE DRAINAGE.

monolith A large hollow rectangular foundation structure, usually of concrete, with several wells or compartments. It is lowered through water

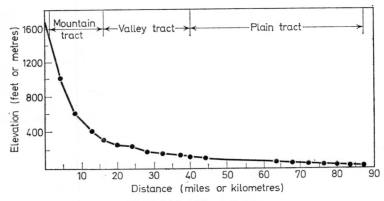

Figure M.3. Mountain tract

and ground by grabbing crane and sealed around the periphery when solid rock is reached. See OPEN CAISSON.

monsoon A wind blowing with regularity and persistence during a definite season of the year; applied especially in Indian Ocean, blowing from NE. in winter (dry monsoon) and from SW. in summer (wet monsoon); a rainy season with winds.

moraines The rock and soil debris which falls on to the surface of a glacier from the valley slopes. It forms two marginal layers or ridges, termed LATERAL MORAINES, parallel to the sides of the glacier. In narrow valleys, the glacier surface is often completely covered by slide or scree material. See MEDIAL MORAINE.

morass A bog or marsh; an area of soft, wet ground. See MARSH, SWAMP.

mosaic Applied to a map composed of a number of AERIAL PHOTOGRAPHS which have been enlarged to the same scale and fitted together. It may be either uncontrolled or controlled. An uncontrolled mosaic is cheaper to produce but it cannot be used to give accurate measurements because it is not related to any ground survey control. On the other hand, a controlled mosaic is more costly but gives detailed features with greater accuracy.

motor truck A self-propelled road vehicle for transporting earth, soil, rock, goods, or other materials at excavations, earthworks, dams, canals, and river improvement sites. Capacities range from 5 to 40 t and for smooth operation adverse road grades should not exceed about 1 in 7. The truck bodies are usually steel with end, side, or bottom discharge and tilting may be performed by hydraulic cylinder and ram. See DUMP TRUCK, DUMP WAGON.

moulins (or mills) Roughly circular and vertical pits, hollows, or cavities formed in the bed of a glacier by the plunging of water through holes in the ice; nearly vertical holes in a glacier formed by running water. See POTHOLES.

mound breakwater See RUBBLE-MOUND BREAKWATER.

mound, groundwater See GROUNDWATER MOUND.

mound spring A spring around which a mound of mineral matter has been precipitated by the spring water or formed by wind-blown sediments and vegetation.

mountain glacier A glacier occupying a valley between mountains which rise above the SNOW LINE. See GLACIERETS.

mountain tract The stretch of a stream along its source in the mountains. The water is swiftly flowing and has carved a steep V shaped depression. In this region, a stream is active in draining the valley slopes and in the erosion and transportation of materials and even flood waters are easily constrained by the steep gorge section (Figure M.3). See PLAIN TRACT, VALLEY TRACT.

movable bridge A bridge which can be moved bodily to permit the passage of ships. See BASCULE BRIDGE, SWING BRIDGE, TRAVERSING BRIDGE.

movable dam A barrier which can be removed wholly or in part to allow increased flow during floods. The movable sections may consists of stop logs, gates, or any other appliance whereby the area for flow may be increased or controlled. See FLASHBOARD.

muck Dirt, earth, or waste material. See BACK CUTTING, BACK FILLING, FILLING.

mud A fine-grained, soft earthy mass with a high content of water; the dominant grain size is less than 0·01 mm. It may consist of sand, clay, limestone, or a mixture of these and other materials in any proportion. See SLIME.

mud flat (or **estuarine flat**) The area of floor of a bay exposed at low tide.

mudflow The downhill movement of waterlogged sediments and debris; may result from runoff after heavy rains. See LANDSLIDE.

Mulberry harbour A prefabricated harbour used as a bridgehead by the Allies in the Normandy landings during June 1944. The precast concrete sections were made in the U.K. and then towed to the Normandy coast of France and there assembled and sunk offshore. See BAILEY BRIDGE.

mulching The prevention or reduction of evaporation from soil by spreading a protective layer of small stones, paper, straw, or plant residue on the surface of the soil.

multiple-arch-type dam A lightweight dam sometimes adopted where foundations are weak; consists of a series of arches supported by parallel buttress walls or piers. The arches transfer the load to the foundations through the buttresses. The Australian civil engineer J. D. Derry carried out pioneer work on the modern type of multiple-arch dam. See COUNTER-ARCHED REVETMENT, FLAT-SLAB DECK DAM, ROUND-HEAD BUTTRESS DAM.

multiple-dome dam A development of the MULTIPLE-ARCH-TYPE DAM consisting of domes supported by buttresses. The 76 m (250 ft) high Collidge dam in Arizona, U.S.A., is of this type.

multiple-purpose reservoir A reservoir constructed to serve two or more purposes, such as water supply, flood control, irrigation, navigation, recreation, pollution control, and perhaps power purposes. See SERVICE RESERVOIR.

multi-storage system A system of water conservation comprising the use of several reservoirs and associated works. Each unit is planned to operate with the others to achieve maximum economy and efficiency in water use. See SIMPLE STORAGE SYSTEM, WATER-RESOURCE SYSTEM.

multi-wheel roller See PNEUMATIC-TYRED ROLLER.

N

nail-bournes Temporary streams. See BOURNES.

nappe (or **sheet**) The sheet of water flowing over the crest of a wall, dam, or weir erected across a stream; has an upper and a lower surface.

narrow-base terrace Similar to a BROAD-BASE TERRACE except that the base width is normally from 1·2 to 2·4 m (4 to 8 ft).

narrows The narrow channel cut by a stream across a highly inclined or vertical hard bed. Since the resistant rock is roughly vertical, the outcrop does not shift as erosion in the narrows proceeds. The stream is considerably wider in the softer beds on each side of the narrows. See RAPIDS, WATERFALLS.

natural escape   See BY-WASH.

natural harbour   A sheltered area of sea formed by a favourable form of coastline. See ARTIFICIAL HARBOUR, HARBOUR OF REFUGE.

natural levee   (Figure N.1) See LEVEE.

natural load   The material carried or transported by a stable stream. See LOAD.

natural recharge   The natural infiltration of surface water or rainfall into underground rocks or aquifers. See ARTIFICIAL RECHARGE.

natural reservoirs   Lakes which lie at or near the base of mountainous regions and from which rivers flow. They reduce the damage from floods

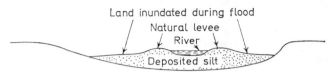

Figure N.1.   Natural levee

and also increase the navigable season of the river. A good example is the chain of Great Lakes, which exercise a large measure of control over the St Lawrence River. See Great Lakes Commission (Appendix).

natural slope   The maximum angle at which rock or soil banks will stand without slipping; the steepest slope of the land. See ANGLE OF REPOSE.

natural storages   Natural features which tend to smooth out the irregularity of stream flow. Natural storages may be underground or on the surface. Marshes, ponds, lakes, and any depressions below flood level usually provide temporary flood storage. Even a thick cover of vegetation will retard and hold part of any excessive runoff. See ARTIFICIAL STORAGES.

natural-strainer well   A well which obtains its water from a partly incoherent deposit and whose SPECIFIC CAPACITY has been increased by the removal of some of the fine particles around the entry section of the well. See GRAVEL FILTER WELL.

natural water   The water which falls as rain, hail, or snow, is derived from dew, or is found in rivers, springs, and surface or underground accumulations. It contains solid, liquid, and gaseous substances, which it holds in solution or suspension. See FRESH, METEORIC WATER, WATER.

natural well   A natural, roughly rounded cavity in the ground which extends downwards into a subsurface body of water. The water does not flow to the surface. See KARST TOPOGRAPHY.

navigation   Methods of fixing a ship's position and course by nautical astronomy, etc. Sometimes applied to a river, the flow of which is controlled by CANALISATION. See SLACK-WATER NAVIGATION.

neap range   See MEAN TIDAL RANGE, NEAP TIDES.

neap tides   The low or smallest tides that occur in the second and fourth quarters of the moon. See SPRING TIDES.

needle A heavy timber piece set either vertically or inclined in a stream to close an opening or to control the flow. See FLASHBOARD, NEEDLE WEIR.

needle valve A valve used in regulating the flow to large WATER TURBINES; cone-shaped and terminating in a sharp point.

needle weir A fixed FRAMED DAM built up by heavy vertical timbers, or needles, in contact. The upstream water level may be lowered by removing the timbers as required.

negative artesian head See ARTESIAN HEAD, NEGATIVE.

negative confining bed A relatively impervious bed below a water-bearing deposit and the downward movement of the water is prevented or retarded. See PERCHED WATER.

net duty (of water) (or **farm duty** (U.S.A.)) The quantity of irrigation water delivered to a farm or irrigation holding; usually expressed in acre feet or cubic metres. See IRRIGATION REQUIREMENT.

net intake A term used by F. H. Edmunds for the amount of water which reaches the zone of saturation. It equals the excess of rainfall over runoff, evaporation, and transpiration.

net peak flow Total peak flow minus the corresponding BASE FLOW or sustained flow in a stream or channel.

net supply interval The rainfall period of a storm between INITIAL RAIN and RESIDUAL RAIN.

neutral pressure The HYDROSTATIC PRESSURE in the water occupying the pore space of a soil. See PORE WATER PRESSURE.

neutral shoreline A shoreline whose essential features do not depend on either the submergence of a former land mass or the emergence of a former subaqueous surface. See COMPOUND COAST.

neutral soil Precisely, a soil with a pH VALUE of 7·0; or, broadly, a soil with a pH value between 6·6 and 7·3.

névé Snow in a state intermediate between FIRN (loose surface snow) and crystalline glacier ice; the snow and ice that creeps down mountain slopes and forms the source of glaciers.

Newlyn datum The MEAN SEA LEVEL at Newlyn, Cornwall (U.K.). See ORDNANCE DATUM.

Nichols terrace A small embankment of earth with a gentle slope on the downhill side, a moderately deep narrow channel, and a flat low ridge. See TERRACES.

nomogram (or **alignment chart**) A chart comprising a set of scales for shortening or eliminating calculations. A frequent form consists of three straight lines each graduated to represent one of the variables in an equation. A set square or straight edge joining values on two of the scales will intersect the required value on the third scale. For example, a simple nomogram may be constructed to represent the values in the equation, discharge = velocity × sectional area. Similar units are used throughout.

non-artesian well A well in which the water is not under pressure and its level in the well roughly coincides with that of the local water-table. See FREE GROUNDWATER.

non-capillary porosity See TOTAL POROSITY.

non-carbonate hardness The term which has replaced the earlier PERMANENT HARDNESS. See HARDNESS OF WATER.

non-contributing area See CLOSED DRAINAGE.

non-dimensional unit graph See DIMENSIONLESS UNIT GRAPH.

non-effective storage See EFFECTIVE STORAGE.

non-flowing artesian well A well which has tapped water with sufficient pressure to cause it to rise in the well but not sufficient to reach and overflow at the surface.

non-frontal precipitation Rainfall which may occur in an area of low pressure; horizontal convergence and inflow into DEPRESSION causes a lifting or upward movement of air. See FRONTAL PRECIPITATION.

non-irrigable land Land which is too high to allow efficient irrigation by gravity means from existing supply channels. Land less than 150 mm (6 in) below the supply level at the farm boundary is regarded by some authorities as non-irrigable by gravity means. See IRRIGABLE LAND.

non-return valve See CLACK VALVE.

non-steady flow See TURBULENT FLOW.

non-storage routing A FLOOD ROUTING method using empirical relationships, such as the water levels at the inflow and outflow ends of the reach considered. See STORAGE ROUTING.

non-tidal river A river beyond the influence of tides. See TIDAL CURRENT.

non-uniform flow A flow which is not constant and the mean velocity varies along the length of the channel. See TURBULENT FLOW.

normal aeration The entire renewal of soil air about once each hour to a depth of 200 mm.

normal cycle (erosion) A CYCLE OF EROSION in which water is the principal erosive agent. See EROSION CLASS, GEOLOGICAL EROSION.

normal depletion curve A curve showing normal loss of water from groundwater storage. See GROUNDWATER DISCHARGE.

normal depth The depth at a given point in a river or open channel corresponding to a steady discharge.

normal erosion See GEOLOGICAL EROSION.

normal flow The volume of water flowing in a uniform length of stream or channel when the water surface slope and bed slope are parallel.

normal recession curve See HYDROGRAPH, RECESSION.

normal soil erosion See GEOLOGICAL EROSION.

normal velocity of distribution The distribution of velocities attained in a straight and uniform length of river or channel between two specified points or cross sections.

normal water level The water level which for 50% of the time during a year or a number of years is equalled or exceeded.

Norton tube well See ABYSSINIAN TUBE WELL.

nose of groyne See GROYNES.

notch The opening in a MEASURING WEIR and can be of rectangular, triangular, or trapezoidal shape.

notched weir See MEASURING WEIR.

notch, triangular See TRIANGULAR NOTCH.

notch, vee See TRIANGULAR NOTCH.

nullah A normally dry watercourse which becomes full and swiftly flowing after heavy rain (India). See FLASHY STREAM.

nun buoy A circular BUOY tapered at the upper and lower ends. See WHISTLING BUOY.

N-year flood A flood of a magnitude which, on the average, once in a given *N* number of years, is equalled or exceeded. See FLOOD (RETURN PERIOD), RETURN PERIOD.

O

obliteration of lakes Ponds and lakes may naturally disappear by (1) enlargement or cutting down of outlet; (2) evaporation, particularly in arid regions; (3) filling of basin with silt, plant growth or both; (4) lowering of surrounding water-table, and (5) a combination of two or more of the above causes. See PERIODIC TYPE (lakes).

O.B.M. See ORDNANCE BENCH MARK.

observation well A well sunk to assist in groundwater investigations and studies. Checks are maintained on groundwater fluctuations and water quality, etc.

occlusion See FRONTAL SURFACE.

ocean A vast expanse of salt water which covers the Earth's crust.

ocean currents Well-defined flows of water over certain portions of the oceans; caused by differences in density of the water, slope of the sea surface, and wind friction. All ocean movements are influenced by the deflective forces caused by the rotation of the earth, etc.

ocean engineering The use of engineering and scientific knowledge of the oceans for the benefit of mankind; would include COASTAL ENGINEERING.

oceanography The science of the oceans; a discourse or treatise dealing with the oceans. It would cover the mapping of the ocean floor and charting the currents and tides leading to better shipping routes, better weather forecasting, and a better understanding of the mineral wealth on and under the sea bed. See THALASSOGRAPHY.

oceanology The science of the oceans. See OCEANOGRAPHY.

O.D. See ORDNANCE DATUM.

off-highway truck In the U.K., a truck constructed to certain standards, with larger tyres, thicker panelling, greater ruggedness, and less chromium trim, etc., than are common for 'on-highway' trucks. See DUMP WAGON, MOTOR TRUCK.

160

offshore terrace A deposit of sand and silt built out into deep water by the combined actions of waves and currents; the deposit is roughly parallel to the shore. See BEACHES.

offshore zone The zone beyond the low tide level consisting of deposits of land-derived sediment forming a terrace with a seaward slope. See SHORE.

offstream storage A reservoir located some distance away from a stream, usually on river flats. The reservoir is filled either by pumping or by diverting water into it by channel or other means.

old river A river with a low slope, a sluggish current, a wide floodplain, and perhaps many oxbow lakes; a river which meanders widely, it erodes during floods and deposits material at other times. See YOUTHFUL STREAM.

on-farm storage See FARM RESERVOIR.

onstream storage A reservoir located on a river or watercourse which is filled by the gravity runoff from the stream CATCHMENT AREA.

open caisson A CAISSON which may be cylindrical, open at the bottom and top, and constructed from steel plate, brickwork, mass concrete or concrete blocks. It is sunk in position to form part of the foundation structure. See MONOLITH, SHIP CAISSON.

open channel Any channel or conduit within which water flows with a free surface. Typical open channels are streams and canals which are UNCOVERED OPEN CHANNELS, while pipes and tunnels when flowing partly full are called COVERED OPEN CHANNELS.

open-end well A well with a pipe form of lining where the water enters near the bottom through a screen or other arrangement in the lining. See PERFORATED-CASING WELL.

open pit (or **open well** or **large bore**) A drilled hole or excavated pit at least 0·9 m (3 ft) in diameter or width which a man can descend to inspect the ground, usually to the water-table or lower if pumping is employed. See DAM SITE INVESTIGATION.

open well Usually an excavated well with a width or diameter of at least 0·9 m (3 ft) and large enough to enable a man to descend to work or inspect the ground possibly down to the water level. *Open hole* is sometimes used to describe a borehole which is drilled without a CORE. See COMBINATION WELL.

open wharf See WHARF.

operational log See DRILLER'S LOG.

operation waste The quantity of water lost from an irrigation district due to the operation of the channel system and includes water passed to waste through outfalls or similar structures. See IRRIGATION, IRRIGATION REQUIREMENT.

optimum moisture content The moisture content of a soil at which a given amount of compaction produces the highest value for DRY DENSITY.

Ordnance bench mark (O.B.M.) A BENCH MARK established by the Ordnance Survey (U.K.) with reference to the MEAN SEA LEVEL at Newlyn, Cornwall. See ORDNANCE DATUM.

Ordnance datum (O.D.) The base or datum used on British Ordnance Survey maps for land altitudes, levels, or levelling work. At an earlier period, it was the assumed mean sea level at Liverpool which is approximately 0·2 m (0·65 ft) below the general mean level of the sea. More recently, the mean sea level at Newlyn, Cornwall, has been used and all revised editions of O.S. maps are referred to this datum. See GENERAL BASE LEVEL OF EROSION.

orifice An aperture or opening in a plate, partition, or wall and generally used for the measurement or control of water. See MEASURING WEIR.

orifice, effective area See EFFECTIVE AREA OF AN ORIFICE.

orifice meter A device for measuring the flow of water; consists of an opening in a plate which is fixed within the pipe carrying the water.

original consequent lakes Lakes formed in original hollows or basins in the land surface, such as depressions in glacial till, in lava flows or in sand dunes. The depressions may have been in the ocean floor and preserved when it was elevated above sea level. See SINK-HOLE LAKES.

orographic precipitation Rainfall caused by the active cooling of air, as when a current is forced upwards over a range of mountains.

oscillation limits See LIMITS OF OSCILLATION.

ounce, avoirdupois Equals 28·3 grammes; 437·5 grains; 1/16 lb.

ounce, fluid Equals 28·41 cm³; 8 fluid drachms.

ounce, troy Equals 31·1 grammes; 480 grains; 1/12 lb.

outburst bank The mid-portion of a sea bank or ridge between the uppermost part, or SWASH BANK, and the footing.

outcrop That part of a rock or other deposit which appears at the surface. Outcropping permeable rocks are important as paths to feed underground aquifers. See INTAKE AREA OF AQUIFER.

outcrop water Surface water which seeps into and along outcropping porous and fissured rocks; water which passes downwards into tunnels, deep excavations, wells, and underground storage.

outfall (1) The site where water is discharged from a conduit into the sea or stream. (2) An irrigation structure used to pass waste water into a drain.

outlet (river) Any artificial or natural escape channel which branches off a river and delivers water into depressions, drainage areas, or into the sea; any artificial or natural escape channel through or over a river bank or levee.

outlet capacity Discharge at the outlet when the main channel is flowing at authorised full supply discharge.

outlet channel Any artificial channel for conveying water from prepared structures such as TERRACES and DIVERSIONS. See ESCAPE.

outlet pipe (small dams) A heavy-duty cast-iron or steel pipe placed along the base of a small dam from the upstream to the downstream side. The pipe flow is controlled by a gate valve which is normally fixed at the downstream end of the pipe. Strong pipes are used to enable them to carry the heavy earth-moving machines moving over them. The pipes are fitted with square mild steel or reinforced concrete CUT-OFF COLLARS (or anti-seep diaphragms) (Figure O.1).

outlet works The works that permit water to be released under control from a reservoir or other storage; includes the conduits, valves, screens, etc.

outwash fan Accumulations of sand and gravel deposited by streams along the front of a glacier. Streams of melt-water often form along the lower

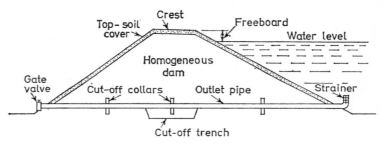

Figure O.1. Outlet pipe (small dams)

part of a glacier and these transport many subangular fragments which are deposited at the ice-front where the water velocity is checked. See ENGLACIAL MATERIAL, ESKERS, KAMES.

outwash plain See VALLEY TRAIN.

overdevelopment See OVERPUMPING.

overfall The crest or part of a dam or weir over which the water flows; the water which overflows.

overfall dam A dam constructed across a stream which allows water to overflow along its entire crest.

overfall weir A low-level weir which permits water to flow over its entire crest.

overflow spillway A structure used on a concrete dam to pass overflow. The dam itself provides the overflow and also the overflow channel. A middle section of the dam forms the overflow and the water passes down the profile of the dam to the river below. On the downstream base of a large dam, precautions are taken to prevent any scouring by overflow water. See APRON.

overflow spring A type of STRATUM SPRING in which the permeable rock dips beneath an impermeable cover. The spring forms an 'overflow' of groundwater at the lower edge of the impervious deposit. Overflow springs often occur along the Tertiary clays which overlie the water-bearing Chalk at Surrey (U.K.).

overflow stand pipe A STAND-PIPE in which the water overflows at the HYDRAULIC GRADIENT or the HYDRAULIC GRADE LINE.

overflow weir   A structure which passes overflow from an earth dam reservoir into a channel or spillway leading to the river below the dam; usually constructed at one extremity of the dam and built upwards from solid ground and not from the embankments.

overhead irrigation   See SPRINKLER IRRIGATION.

overland flow   That part of storm water which flows over the ground before it cuts its own channel or reaches a stream. See SPILL, SURFACE RUNOFF.

overland runoff   See SURFACE RUNOFF.

overloader   A POWER SHOVEL type of loading machine which may be either pneumatic-tyred or caterpillar-tracked. The machine need not turn to load into a truck if the latter can be positioned alongside. The bucket is filled, the machine retracted, and the bucket swung over to the truck or discharge point. Used for excavating and loading earth, soil, or other material, sometimes in BORROW PITS. See MOTOR TRUCK.

overnight pond   A small pond connected to an irrigation system for impounding water and from which supplies may be drawn when required. See PONDAGE, STORAGE.

overpumping (or **over development**)   Applied to a well from which the water pumped out exceeds the rate of replenishment from the outcropping parts of the aquifer. See ECONOMIC YIELD, YIELD.

overshot spillway   A spillway consisting of a flume formed directly over the dam or embankment. See DROP-INLET SPILLWAY.

oversize cobbles (earth dams)   Cobbles too large to be placed in earth dams. When a sheepsfoot roller is used for compaction, the maximum diameter should be between 125 and 150 mm (5 and 6 in).

oxbow   A bend in a river; the extreme curvatures of a meandering river. See FULL MEANDER.

oxbow lake   A lake formed in a meander area which has been abandoned by the stream cutting across the narrow neck of the loop. See INGROWN MEANDERS, MEANDER BELT.

ozonisation   A method which utilises the powerful oxidising action of the gas, O_3, to sterilise water supplies and also to produce high-quality water for mineral water factories and for breweries. The ozone may be brought into intimate contact with the water either by blowing it through a distribution system within the water or by means of an injector. See CHLORINATION.

P

Pacific type (coastline)   A coast which runs parallel to the main structural lines of the adjoining land masses. See ATLANTIC TYPE, MODIFIED PACIFIC TYPE.

packing   Hemp or similar material used in a STUFFING BOX to prevent leakage of water from a pump.

paddle   A wooden panel used in a lock or sluice or other hydraulic channel to close a water passage. See LOCK PADDLE.

paddle hole   An opening which allows the entry or discharge of water from a canal LOCK or other hydraulic structure.

padudal   Pertaining to swamps and marshes. See FLUVIAL, LIMNOLOGY.

pan   See HARDPAN.

Panama Canal   A canal connecting the Atlantic and Pacific Oceans; 80·5 km (50 miles) long (with sea approaches) and between 12·5 and 26 m (41 and 85 ft) deep. The construction of the canal was hindered greatly by LANDSLIDES. Large slides in the softer rocks totalling about 38 million m^3 (50 million yd^3) had to be cleared in addition to the canal excavation. See SUEZ CANAL.

pan coefficient   Ratio of evaporation from a reservoir, lake, or other body of water to that given by an evaporation pan. See EVAPORIMETER.

panel   The surface area of water enclosed between the corresponding traces of SEGMENTS in adjacent cross-sections of a stream.

panplain   A very broad plain formed by the union of adjacent river valleys. It has a planed-off bedrock floor covered with a thin deposit of alluvium. See FLOODPLAIN, PENEPLAIN, VALLEY FLOODPLAIN.

parapet wall (or storm wall or wave wall)   A wall often built on the up-stream side of the roadway or pathway formed along the top width of a dam.

Parshall measuring flume   A calibrated device or flume developed by the U.S. Department of Agriculture to measure flow of water in open conduits; consists essentially of a contracting upstream section, a throat, and an expanding downstream section. The water is intended to flow at BELANGER'S CRITICAL DEPTH over a sill placed at the throat. Formerly called IMPROVED VENTURI FLUME.

partial diversion (or partial cutoff)   A cutoff in a river where total diversion of the current is not desirable. The cut is of sufficient size to take large quantities of flood water, but the river still follows the old course with a reduced flow. A partial diversion may be adopted where property owners on the old course depend on the river for water supply. See DIVERSION.

partial drought   In the U.K. a period of at least 29 consecutive days, the mean daily rainfall of which does not exceed 0·25 mm (0·01 in) *(British Rainfall, 1887)*.

partial duration series   A list of hydrologic events, within a specific period, occurring above a selected base and without regard to number. The selected base, in the case of floods, is generally equal to the lowest annual flood to enable at least one flood being included in each year. In contrast, a tabulation or graph in which only the maximum peak for each year is included in the series is termed ANNUAL FLOOD PEAK SERIES.

partially separate system   A drainage method in which the rainwater from backyards and the roofs of dwellings flows away with the dwelling sewage and a different sewer collects the rest of the rainfall. See COMBINED SYSTEM, SEPARATE SYSTEM.

Pascal, Blaise (1623–62) A French mathematician who invented the HYDROSTATIC PRESS. See PASCAL'S LAW.

Pascal's law A law of fluid pressure (1646) which states that pressure exerted anywhere to an enclosed body of fluid is transmitted undiminished in all directions. This pressure per unit area is uniform throughout and acts at right angles to every part of the surface of the container. See HYDROSTATIC PRESS.

passive method A construction method sometimes used to protect buildings erected in PERMAFROST areas. The frozen ground is left intact and the undersides of buildings are well insulated to prevent heat from the structures thawing the underlying ground. See ACTIVE METHOD.

path of percolation See LINE OF CREEP.

paving, stone See STONE PAVING.

peak flood The point and time of maximum flow during a specific flood. See FLOOD.

peak flow, net See NET PEAK FLOW.

peak rate of runoff The maximum rate at which water flows from a catchment area during a given storm; usually expressed in cumecs or cusecs. See MAXIMUM RAINFALL INTENSITY.

peak unit flow The maximum or peak of a UNIT HYDROGRAPH.

peat A surface covering of more or less decomposed vegetable matter, usually with a high water content. Where peat exists within a reservoir area it may cause acidity and colouring of the water. The remedies include removal of the peat, wholly or in part, or covering the deposit with clean soil from nearby areas.

pegging out See STAKING

pellicular water See ADHESIVE WATER.

pellicular zone The ground or depth below the surface to which evaporation effects can penetrate.

Pelton water wheel An impulse type of water wheel which obtains its rotative force from the impulse of a powerful jet of water against buckets spaced at its periphery. The wheel requires a head of at least 275 m (900 ft) for efficient operation. See IMPULSE TURBINE.

pendent A small metallic disc or appendage, carried by a PENDENT WIRE, and stamped with an observation point number.

pendent wire A wire marking the direction of a section line and along which the position of observation points are marked by PENDENTS.

peneplain (sometimes **peneplane**) A featureless, roughly flat land area, at or near sea level and formed by long-continued erosion and weathering. 'Peneplain' (L. *paene*, almost) was coined by William Morris Davis.

penning gate A rectangular sluice gate which moves upwards to open. Used chiefly in U.K. See LIFT GATE.

penstock A pipe which conveys water under pressure to a WATER TURBINE.

percent saturation See DEGREE OF SATURATION.

perched stream An INSULATED STREAM or a stream perched well above the zone of saturation.

perched water A body of subsurface water which is separated from the underlying zone of saturation by a confining bed or impervious stratum.

perched water-table The upper limit or surface of a small body of water above the MAIN WATER-TABLE. The water is retained in its elevated position by an impervious stratum and may form a limited source of water supply.

percolating filter See BACTERIA BED.

percolation That part of rainfall that enters the soil and subsoil. The water that moves downwards and laterally into streams is termed SHALLOW PERCOLATION, while the water that continues downwards into the zone of saturation or other subsurface accumulation is termed DEEP PERCOLATION. In general, percolation is a maximum in winter and almost nil in summer.

percolation gauge A cylinder or block of earth surrounded by brick, concrete, or other material, and so set that water from rainfall may be collected at the base of the cylinder. Gauges vary in depth from 0·5 to 1·5 m (20 to 60 in) and in area from 0·8 to 4 m² (1 to 5 yd²). They may be left bare or turfed. The INSTITUTION OF WATER ENGINEERS (Appendix) has produced a specification for percolation gauges. Records of percolation from gauges are published annually in *British Rainfall*. See INFILTRO-METER.

percolation path (dams) The direction or path taken by water percolating beneath a dam from the upstream to the downstream side. It is both a source of leakage from the reservoir and also liable to exert an upward pressure on the base of the dam. When it occurs in permeable foundation rocks a reduction is possible by making the percolation path as long as possible, which lowers the hydraulic gradient between the upstream and downstream faces of the dam. This could be done by a cutoff trench, packed with impervious material to a designed depth and constructed along the length of the foundation near the upstream face. This impervious barrier deflects the percolation downwards and increases its path to the downstream side. The cutoff may consist of a vertical wall of grouted rock or a curtain of sheet piling. Where a dam is to be constructed on porous deposits, a horizontal concrete apron may be formed, extending some distance upstream and downstream from the dam. This construction also increases the length of the percolation path to the downstream side. See PIPING, TUNNELLING.

percolation rate The percolation per unit of time.

perennial snowfields Cold regions, such as polar lands and some high mountains, where the snowfall remains throughout the year. See SNOW LINE.

perennial spring A SPRING in which the flow of water is unceasing. See INTERMITTENT SPRING.

perennial stream A stream that flows continuously throughout the year and often stands lower than the local water-table. See EFFLUENT SEEPAGE.

perforated-casing well A well which the water enters through holes in the casing. Commonly applicable to steel or iron cased wells, but also includes wells with different types of casing. See OPEN-END WELL.

periglacial The ground or zone peripheral to a glacier or ice sheet; phenomena associated with the Glacial age. See PERMAFROST.

perimeter, wetted See WETTED PERIMETER.

periodic spring A type of INTERMITTENT SPRING in which the discharge is relatively great at roughly regular and frequent intervals. It often occurs in limestone districts and its rhythmic discharge may be due to a natural siphon underground. See BREATHING WELL.

periodic type (lakes) Some lakes found in arid regions, without outlet, which evaporate to dryness during a portion of the year. See LAKE-FLOOR PLAIN, OBLITERATION OF LAKES, PLAYA LAKES.

period of concentration (storm) The storm duration which allows the rainfall on substantially the entire catchment area to have its effect on the flow at the outfall. Sometimes called the RESPONSE TIME. See FLOOD HYDROGRAPH.

period of pulsation The average period of a cycle of pulsation at a point in the cross-section of a stream during which the velocity fluctuates between limiting low and high values. See MEAN VELOCITY ON A VERTICAL.

permafrost Ground perennially frozen; occurs in regions where the mean annual temperature is below freezing point. It forms a near-surface deposit which is impervious and may be hundreds of feet in thickness. In areas of permafrost certain engineering problems may arise, such as contamination of surface water supplies, disposal of sewage from large camps and foundation work, etc. The term was coined by S.W. Muller (1945, *U.S.G.S. Special Report*). See ACTIVE METHOD.

permafrost table The upper surface or limit of PERMAFROST.

permanent bank protection Permanent measures or works to check erosion along river banks. They may consist of concrete or masonry walls, concrete blocks, or concrete sheet piling, or broken rock tipped from the bank to form a continuous apron. Such works are costly and are used only where important structures, such as bridges or buildings, require protection. They are often used within towns where bank erosion must be arrested. Strong-rooting, quick-growing plants may be viewed as permanent protection. In New Zealand and Australia, poplars and willows are sometimes used because they grow rapidly under a wide range of conditions. See BANK PROTECTION, STONE RIPRAP.

permanent hardness (or **non-carbonate hardness**) See HARDNESS OF WATER.

permanent stream A PERENNIAL STREAM or one with an unceasing flow. All streams positioned below the water-table are usually permanent. See TEMPORARY STREAM.

permanent wilting point The stage when a soil is so dry that plant roots are unable to extract water at a rate sufficient to maintain turgidity. At this stage, transpiration almost ceases, the plants wilt, and the soil is said to be at permanent wilting point. Temporary wilting may occur on a hot day, but in this case the soil moisture may be plentiful. See WILTING COEFFICIENT.

permeability The capacity of a rock, soil, or other substance for transmitting water. Shales and clays are relatively impermeable, while sands, gravels, and sandstones with their much higher pore space and inter-communication are permeable. Impervious rocks are important in the case of reservoirs and earth dams for ponds, while permeable soils are desirable in the case of roadways and railways. See HAZEN'S LAW.

permeability coefficient See COEFFICIENT OF PERMEABILITY.

permeability, field coefficient of See FIELD COEFFICIENT OF PERMEABILITY.

permeable groynes (or **pile and waling groynes**) Usually a timber construction consisting of a double row of piles strutted together. Walings or horizontal timber planks are fixed to the upstream row of piles and spaced about their own width. Sometimes old steel rails are used as piles. Permeable groynes which are pointed upstream tend to collect much floating debris and so lose permeability. The pile and waling type can be built out stage by stage. See GROYNES, IMPERMEABLE GROYNES.

permeameter A laboratory instrument for measuring the COEFFICIENT OF PERMEABILITY of a soil sample. The falling-head permeameter is used for impermeable soils such as fine silts and clays and the constant-head type is used for permeable gravel or sand.

permissible velocity (or **safe velocity**) The maximum velocity of flow of water that can exist in a channel and not cause erosion; depends on the nature of the ground and other factors; the highest velocity at which water may be conveyed safely in a conduit.

pervious rock A rock that permits the passage of water with relative ease, as opposed to an impervious rock. See MICROPORES.

Pettis's formula A flood formula as follows:

$$Q = CRb^{1 \cdot 25}$$

where R is a rainfall coefficient, b is the average width of the catchment, C is a coefficient to cover other factors. (*Engineering News Record*, June 21, 1924). See FLOOD FORMULAS.

pF value For held water in soil, the pF value is equal to the common logarithm of the suction given in centimetres of water. See SOIL MOISTURE SUCTION.

pH value A convenient expression for the acidity or alkalinity of a solution; is the logarithm of the reciprocal of the hydrogen ion concentration. A one-unit change in pH value is equivalent to a tenfold change in hydrogen ion concentration. Distilled water has a pH value of 7, the neutral point; all values above 7 indicate the presence of alkalis, and all below 7 indicate acids. See ACID WATER.

photogeology   The study and geological interpretation of vertical aerial photographs. Of maximum value in open country where rock formations are partially exposed. Even aerial photographs of an area covered by dense vegetation often disclose the main geological features, particularly where colour variations of vegetation can be discerned and correlated to the nature of the soils and subsurface rocks. See INFRA-RED PHOTOGRAPHY, MOSAIC.

photo-grammetrist   A person skilled in preparing maps from aerial photographs and in the study of overlapping photographs in pairs under a stereoscope. He can prepare both topographical and structure-contour maps.

phreatic   A term not in general use but sometimes applied to groundwater and associated phenomena.

phreatic cycle   See CYCLE OF FLUCTUATION.

phreatic decline   The depression or downward movement of the water-table. See DRAWDOWN.

phreatic divide   See GROUNDWATER DIVIDE.

phreatic fluctuation   See FLUCTUATION OF WATER-TABLE.

phreatic fluctuation, belt of   See BELT OF PHREATIC FLUCTUATION.

phreatic high   The highest level reached by the water-table at a given station and for a given PHREATIC CYCLE. Similarly, the PHREATIC LOW is the lowest level reached by the water-table. Usually expressed in metres or feet above mean sea level or below a datum on the land surface.

phreatic low   See PHREATIC HIGH.

phreatic surface   See MAIN WATER-TABLE.

phreatic water   GROUNDWATER; water in the ZONE OF SATURATION; a term not often used in the U.K.

phreatic water discharge   See GROUNDWATER DECREMENT.

phreatic wave (or wave of the water-table)   A swell, undulation, or vibration in the water-table that moves laterally away from the zone where a large intake of water occurs in the zone of saturation.

phreatophyte   A plant that habitually obtains its water supply either through the capillary fringe or directly from the zone of saturation.

physical geology   The science which deals with the activity of natural agents on the earth's surface. See ENGINEER-HYDROLOGIST, ENGINEERING GEOLOGY, GEOLOGY.

phytometer   An appliance to determine transpiration. A sample of soil in which some plants are rooted, is placed in a vessel and so sealed that water can escape only by transpiration from the plants. See EVAPO-TRANSPIRATION.

Piedmont glacier   Formed when two or more valley glaciers unite and spread out on the plains below the mountains; often seen in glacial regions where the SNOW LINE is at a low level.

pier   (1) A circular or rectangular column or wall of plain or reinforced concrete or masonry for carrying heavy loads such as a bridge or other structure. (2) A MOLE or BREAKWATER used as a promenade and landing-stage.

piestic interval The vertical height or distance between two ISOPIESTIC LINES on a map.

piezometer A tube or appliance for measuring PRESSURE HEAD; usually a small pipe tapped into the side of a closed or open conduit and connected to a gauge.

piezometer tubes (dams) Copper or usually plastic tubes inserted in clay fills of dams to measure pore pressure. They are almost standard practice in the U.K., where embankment materials give high pore pressures. As a result of experience in several U.K. earth dams, the British Research Station has developed the 'B.R.S.' apparatus. See PORE WATER PRESSURE.

piezometric surface (or **potentiometric surface**) The imaginary surface to which water will rise under its full head from a given groundwater reservoir.

pile and fascine A method of protecting a steep eroding bank by driving a row of piles at about 3 m (10 ft) intervals, and then packing the space between the piles and the bank with brushwood fascines. The tops of the piles are secured by wire to stakes set well back on the bank. See STAKE AND STONE SAUSAGE CONSTRUCTION.

pile and waling groynes See PERMEABLE GROYNES.

piles Timber, concrete, or steel posts or plates driven into the ground to carry a vertical load or a horizontal load. Steel is usually preferred. Piling is used for foundations and for protective works associated with weirs, aprons, groynes, and general river and reservoir structures.

pillar drains (dams) Pillars consisting of hand-packed stone carried up from the foundations of an earth dam at intervals along its length. They are usually connected at their lower and upper ends by lateral drains. The lower drains are connected to a main drain in which the flow of water can be inspected and measured. See VERTICAL SAND DRAIN.

pilot channels Channels or cuts made along the river outside the noses of groynes to establish the main flow in a new line. This is often necessary if the river bed is heavy and difficult to erode. Where machines are available for this work, the excavated material is often dumped along the groynes and also along the bank between the groynes. Willows are often planted in the dumped material to improve the banks and the stability of the groynes. See GROYNES.

pilot cut-off (or **pilot cut**) A CUT-OFF where erosion is deliberately used to enlarge the cut; only possible where the bed is easily erodible.

pinching of strata The attenuation of strata in a certain direction and finally some beds may 'pinch out' altogether. This structure is important in its effects upon the movement or storage of water within the rocks. A well may not give the expected yield on account of the thinning and perhaps disappearance of the aquifer. See STRUCTURAL TRAP.

pipeline A pipe to convey water over long distances; laid on the surface or in a trench which is then filled in or covered over. See HYDRAULIC TRANSPORT.

pipeline, irrigation A closed conduit for conveying irrigation water. This method is more expensive than unlined earth channels but has the advantage that the CONVEYANCE LOSS is reduced to a minimum.

pipe outlet (farm dam) A pipe set in the base of a small farm dam to pass COMPENSATION WATER.

pipe outlet (irrigation) A pipe placed in the bank of a farm supply ditch. It is fitted with a slide gate to enable water to be discharged from the ditch on to the land. See ROSTER SYSTEM.

piping (or **subsurface erosion**) The erosion caused by seepage of water within and around the base of a structure, dam, or bank and which could affect the safety of the construction. It is caused by water seeping through the dam with a velocity sufficiently strong to carry soil particles. A hole or 'pipe' is created at the downstream face and it gradually moves towards the water face. See BOIL, TUNNELLING.

piracy (river) See RIVER CAPTURE.

piston pump See DIRECT ACTING PUMP, PUMP.

pitched work (or **pitching**) A stone REVETMENT to protect the slopes of a river bank or reservoir from scour. As a protection against heavy waves, the stones are at least 150 mm (6 in) square on the face with a depth of at least 0·5 m (21 in). See BEACHING, REVETMENT.

pitching See PITCHED WORK.

pitometer A device based on the PITOT TUBE principle and used for determining the velocity of flow of fluids in closed conduits and pipes.

Pitot tube A device for measuring the pressure and velocity of flowing

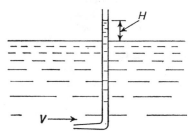

Figure P.1.   Pitot tube

water. In its simplest form (Figure P.1), consists of a tube or pipe bent at right angles with open ends. One arm is shorter than the other, and when the device is placed in a current the water rises in the longer arm, thus

$$H = V^2/2g$$

172

where H is the height of water in longer arm, V is the velocity of water, and g is the acceleration due to gravity.

Other forms consist of two openings (upstream and downstream) and two columns of water. The velocity head is computed from the difference of the two water levels.

pitting The putting down of TRIAL PITS to test and sample shallow alluvial deposits. In soft dry ground, depths of about 15 m (50 ft) may be reached. Pitting may also be used to locate the water-table when at shallow depths and the pits may be used for water supplies. See SHALLOW WELL.

pivot bridge See SWING BRIDGE.

pivot point layout The location of observation points in a river, without direct measurement along the DISCHARGE SECTION LINE, by means of a geometrical layout of points on one or both banks.

plains Land forms whose predominating feature is flatness or low inclination; generally applied to topographic features which cannot properly be described as benches or terraces. Plains of several kinds are formed by rivers. See AGGRADED FLOOD PLAIN, PANPLAIN.

plain tract (or alluvial tract) The stretch of a river approaching its mouth where it flows over a broad flood plain with deposition of fine sediment (Figure M.3). See DELTAIC TRACT, VALLEY TRACT.

plan (or map) (1) A scale drawing, on a horizontal plane, of part of the earth's surface showing natural and artificial features such as geological exposures, faults, folds, excavations, roads, rivers, dams, drainage, contours, and perhaps irrigation and other relevant features. A mine plan would show mine workings, adits, shafts, etc. (2) A map kept by a water undertaker showing the positions of all pipes except service pipes. In the U.K. it may be an Ordnance Survey map or tracing on which all mains, etc., are shown in some distinctive colour. See GEOLOGICAL MAP, INDEX MAP, LITHOLOGICAL MAP, WATER PROSPECT MAP.

planation The action of a river when it cuts a flat-bottomed, trough-shaped valley.

plane of saturation The WATER-TABLE. See MAIN WATER-TABLE.

planimeter An instrument for measuring an area on a plan, whose perimeter has been traced out by its moving arm; available in several units and scales; used in water engineering to compute reservoir area, etc.

planning (impounding schemes) See IMPOUNDING SCHEMES (PLANNING).

planning, waterworks See WATERWORKS PLANNING.

playa lakes Broad, shallow sheets of water in desert regions; they generally form and evaporate after a short period, leaving mud flats or PLAYAS.

playas The level plains of ephemeral lakes in desert regions; deposition of sediment occurs during the inflowing of streams after local cloudbursts and may be followed by a drought of many months' duration. See LAKE-FLOOR PLAIN.

plum A large stone of irregular shape placed in mass concrete dams to reduce the cost of the concrete. See CYCLOPEAN.

plunger See FORCE PUMP, RAM.

plutonic water See JUVENILE WATER, MAGMATIC WATER.

pluvial index The index of total rainfall depth which will probably be exceeded or equalled during a certain period in a specified area over a given number of years.

pluviograph (1) A graph indicating stream flow if all precipitation became SURFACE RUNOFF. (2) A graph showing total rainfall. (3) A self-recording RAIN-GAUGE. See HYDROGRAPH.

pluviometer See RAIN-GAUGE.

pneumatic sewage ejector A DISPLACEMENT PUMP, operated by compressed air, for moving sewage.

pneumatic-tyred roller (or **multi-wheel roller**) A towed roller used for compacting earth dams and similar work. It is carried on two axles, each incorporating several rubber tyres. It is loaded with ballast and may weigh up to 200 t. It compacts the earth in 100 to 150 mm (4 to 6 in) layers. See SUPERFICIAL COMPACTION.

pneumatic water barrel (or **vacuum tank**) An earlier type of WATER BARREL, used mainly for dewatering a sinking shaft, and developed by the late Prof. W. Galloway. By means of a hose connection to an air-pump at the surface, a partial vacuum is created inside the barrel in the shaft. As a result the water lifts the inlet valve and fills the barrel. The hose is then detached and the barrel hoisted to the surface and discharged. Various types of CENTRIFUGAL PUMPS or SUBMERSIBLE PUMPS are now used.

pocket beaches Small shore areas formed at the heads of coves.

point gauge An appliance for measuring water level; consists of a sharp point attached to an arrangement which slides on a graduated staff. The point is lowered until it just touches the free surface of the water. See HOOK GAUGE, STAGE.

point rainfall A rainfall measurement at a single point or station. See RAIN-GAUGE.

point sample (sediment) Taking a sample of SEDIMENT WATER at a single point; a time-integrating sampler or an instantaneous instrument may be used.

poised stream A stream which may be viewed as stable from an engineering aspect, but not possibly from the geologic aspect. See REJUVENATION.

polder An area of marshy and low-lying land reclaimed from the sea and cultivated (Holland). See LAND ACCRETION.

pollard A tree with its top cut off to allow the stump to produce a more dense growth of limbs and branches. The stump is usually about 1·8 m (6 ft) high and this height reduces the tendency to be blown over. 'Pollarding' refers to the practice of pruning and cutting back trees along river banks and other margins.

pond (1) A body of relatively stagnant water, usually smaller than a lake, and often formed artificially by earth ridges or ground excavations. (2) The water retained between two canal locks. See FARM POND, REACH.

pondage (or **balance storage** or **regulator storage**)   The storage of a volume of water, usually only sufficient for one day, during which outflow varies according to usage and inflow may be fairly constant. The term is often applied to water stored temporarily within low banks at a farm; the water is periodically pumped or diverted into the FARM RESERVOIR.

pondage method   A SEEPAGE LOSS determination in an earthen channel. The test section of channel is isolated with temporary water-tight bulkheads and then filled with water. The decrease in volume of water is measured and the seepage is computed with allowances for rainfall and evaporation. See INFLOW-OUTFLOW METHOD.

ponding   Applied to the temporary HEADING UP or rise in water level above a point in a stream.

pond management   The application of practices to prevent siltation and erosion of ponds, fire protection, and stock water and development of uses such as production of fish and perhaps wildlife.

pontoon   A BARGE, flat-bottomed craft, or a light framework or float used in military engineering for bridge building and in civil engineering for general duties on construction projects. See FLOATING HARBOUR.

pontoon bridge   A bridge which is carried on pontoons moored to the river or lake bed. It may be temporary or permanent. Permanent types are sometimes adopted when the foundation material is very soft.

pool spring   A spring that discharges water from a deep pool. Many pool springs are related to faulting structures.

poplars   Large trees of rapid growth with soft wood; propagated by root suckers or from branch cuttings. Their deep root system is often utilised to bind the soil layers on slopes liable to landslides. See TREE PLANTING, WILLOWS.

pore pressure   The pressure of water and air in the VOIDS or INTERSTICES between the grains or particles of a rock or soil mass.

pores   See MACROPORES, MICROPORES, VOIDS.

pore water   Water occupying the interstices, voids, or pores of rocks or soils. See MICROPORES, SATURATION.

pore water pressure   The pressure of water in a saturated soil; may be determined by a BOURDON PRESSURE GAUGE. During the construction of earth dams, the pore water pressure is often measured during and after completion of work because it gives an indication of the degree of compaction. The compaction process is complete if the pressure is zero. See NEUTRAL PRESSURE.

pore water pressure cells   Sensitive instruments which measure PORE WATER PRESSURES produced by load fluctuations such as the rise and fall of a tide.

porosity   The percentage ratio of the volume of voids to the total volume of a rock or soil sample. Thus:

$$P = 100 \frac{W-D}{W-S}$$

where P is the per cent porosity, W is the saturated weight, D is the dry weight, and S is the weight of saturated sample when suspended in water. As a guide, the porosity of sandy soils is greater than 50%; clays 45% or higher; sandstones and limestones up to about 25%, while granites and other igneous rocks are less than 1%. See EFFECTIVE POROSITY, TOTAL POROSITY.

porosity, effective   See EFFECTIVE POROSITY.

porous rock   A rock which contains PORES, and transmits water with relative ease. At a dam site or other storage, a porous bed may be sealed off by grouting or by concreting a trench cut along its outcrop.

portal   The inlet to, or outlet from, a water tunnel; referred to as the upstream or downstream portal, respectively.

Portland cement   A substance which binds AGGREGATES into a hard CONCRETE within a few days; generally made from chalk and clay by mixing in suitable proportions, burning in kilns, and then grinding to a fine powder. Portland cement is stronger than natural cement and sets more slowly. See CONCRETE DAM.

position head   The STATIC HEAD of water.

positive artesian head   See ARTESIAN HEAD, POSITIVE.

positive confining bed   A CONFINING BED which prevents the upward movement of groundwater with a positive artesian head. See ARTESIAN HEAD, POSITIVE.

post-Pliocene   See QUATERNARY.

potable water   Drinkable water; water used for hygiene and amenity purposes. Usually it must comply with statutory requirements, such as those relating to Section 41 of the Factories Act 1937 in the U.K. See INDUSTRIAL WATER.

potamology   A treatise on rivers; branch of hydrology pertaining to streams and rivers. See LIMNOLOGY.

potential (groundwater)   (1) Equals the sum of the PRESSURE HEAD of groundwater at a given point and the elevation of the point above a fixed plane of reference. (2) The groundwater resources available in a region.

potential energy   The energy of water due to its position or height above a given datum. See ENERGY HEAD.

potential gradient (groundwater)   The rate of change in potential in a mass of groundwater. Where no gradient direction is specified, that of maximum gradient is taken.

potential head (or **elevation head**)   The height of any particle of water, in metres or feet, above a datum or point is its elevation head above that datum or point.

potential transpiration   The TRANSPIRATION that would take place if the water for the vegetation were always adequate. In the U.K. estimates of potential transpiration have been made for the summer months April to September from the relevant weather data (air temperature, sunshine, air humidity, and wind speed). When placed alongside the rainfall for the same places and periods, they indicate the frequency of irrigation need.

potential yield   The maximum rate at which water may be extracted from an aquifer throughout the foreseeable future and ignoring the cost of recovery. See ECONOMIC YIELD.

potentiometric surface   See PIEZOMETRIC SURFACE.

potholes   Roughly cylindrical holes cut in the solid rock bed of a stream by the swirling of pebbles or sand particles in eddies. The fragments are rolled at one place and thus bore downwards a roundish hole which can be up to 6 m (20 ft) in diameter and 3 to 4·5 m (10 to 15 ft) deep. See MOULINS.

power earth auger   See EARTH AUGER.

power shovel   A loading machine, usually diesel or electrically driven, with caterpillar mountings or self-laying tracks. It consists of a large shovel or bucket at the end of an arm suspended from a boom which extends crane-like from the machine. When digging, the bucket moves forward and up-ward so that the machine does not usually excavate below the level at which it stands. Used for a wide variety of earth-moving operations. See OVERLOADER.

power take-off   An external splined shaft on a tractor; used while the vehicle is idle for driving water pumps, etc.

precipitable water   The total amount of water produced if all water vapour, in a total vertical column of air of unit horizontal cross section, were con-densed and collected in a receptacle of the same horizontal cross section.

precipitable water, effective   The water vapour in the atmosphere that can be precipitated by efficient storm conditions; expressed as millimetres or inches depth of water.

precipitation   (1) The total amount of water which falls as rain, hail, or snow; expressed as millimetres or inches of rainfall over a specified period; moisture deposited as dew. (2) The settling of solids in sewage or other cleaning or separation process; the chemical process whereby the insoluble substance in a solution settles out. See RAINFALL, RAIN-GAUGE, RATE OF RAINFALL.

precipitation, excessive   A precipitation rate which is greater than certain specified limits selected with regard to the normal rainfall of the area. EXCESSIVE STORMS are storms which cause excessive precipitation.

precipitation mass curve   A curve showing accumulated rainfall plotted against time. See MASS DIAGRAM.

precipitation, trace of   A precipitation rate too low to be measured in a rain-gauge; usually below 0·125 mm (0·005 in).

prescribed rights (water)   Legal title to the use of water acquired by posses-sion and use over a long period without protest of other parties. See WATER RIGHTS.

pressure, effective   See EFFECTIVE PRESSURE.

pressure filter   Generally a cylindrical filter, rather similar to a GRAVITY

FILTER, for removing suspended matter by means of a sand bed. It may be vertical or horizontal, depending on the direction of the axis of the cylinder. The drop in pressure between the unfiltered and filtered water pipes is the HEAD available for flow. See FILTER.

pressure gauge An instrument which measures fluid pressure, such as the BOURDON PRESSURE GAUGE.

pressure grouting See BLANKET GROUTING, CURTAIN GROUTING, GROUTING.

pressure head (or **head**) Describes the water pressure in a system and may be expressed as either a pressure in N/mm^2 (lbf/in^2) or as metres (feet) head.

pressure, impact See IMPACT PRESSURE.

pressure pillow A snow-measuring device consisting of a circular flat bladder about 75 mm (3 in) deep. The pillow is filled with an antifreeze solution and when the snow builds up on it the resulting increased pressure is recorded by an open-end manometer.

pressure-reducing valve A valve designed for regulating or maintaining a reasonably constant pressure. It consists of a weighed piston moving vertically in a cylinder connected by a tube with the main downstream of the valve. It can be made to close at any desired outlet pressure by adjusting the weights on the piston. The valve must be kept in good working order.

pressure-relief valve A safety valve designed to relieve pressure in a pipeline. See AIR VALVE.

pressure sounder A device, based on Boyle's law, for measuring depths.

pressure-sustaining valve A valve which tends to close if the upstream pressure falls and thus a relatively constant upstream pressure is maintained; sometimes used on the inlet to service reservoirs to ensure that the pressure does not fall below a fixed minimum.

pressure tunnel A water supply tunnel in which the water is under considerable pressure. Usually applied to a tunnel driven below some obstruction, such as weak permeable ground or a river bed, and lined with reinforced concrete, cast-iron, or steel and backed with concrete or with broken rock and grouted.

prestressed concrete Concrete which has been stressed before it is required to sustain a load. The procedure greatly reduces or prevents the concrete cracking when in use. Prestressed concrete has been used for water-retaining tanks and its use appears to be economical for reservoirs if the capacity is greater than 2250 m^3 (0·5 m.g.). See CONCRETE, REINFORCED CONCRETE.

prestressed gravity dam A type of GRAVITY DAM in which the resistance to overturning, sliding, or pivoting is provided by its weight and also the dam is anchored to the foundation rock by a system of stressed steel bars or tendons. The design enables the volume of concrete to be reduced by about 40% but this saving is partly offset by the cost of steel rods, anchorages, labour, etc.

preventative snagging (or **stumping off**) The removal of trees growing on the bank of a river and which are liable to fall into the channel and tear down part of the bank with their roots. They are cut down and made to fall landwards. See SNAGGING.

priming (1) The first filling of a reservoir, canal, or other structure with water; may occur each season, or year, or only once during its lifetime. (2) Filling a pump or siphon with water to ensure a full flow at the commencement.

primitive water Sometimes used to designate water trapped in the Earth's interior since its formation; INTERSTITIAL WATER of internal origin. See MAGMATIC WATER.

principle of Archimedes See ARCHIMEDES'S PRINCIPLE.

prior rivers A term often used in Australia to describe river systems which are prior to the present system but postdate the ANCESTRAL RIVERS.

prismoidal formula A formula to determine the volume of an excavated earth water tank, channel, etc. Thus:

$$\text{volume} = \frac{D}{6} (A_1 + A_2 + 4A_m)$$

where D is the depth of excavation,
A_1 and A_2 are the two end areas, and
A_m is the area at the midpoint.

probable maximum floods See FLOOD (return period).

probing (1) Investigating underwater conditions of spur pavements, protection of weirs on upstream and downstream sides, and underwater masonry, etc. (2) Investigating ground conditions by forcing down a pointed steel rod up to about 6 m (20 ft) long. It may be used to locate bedrock, shallow water supplies, etc.

Proctor test Standardised tests for compacting soils in the laboratory to give comparable results. The density of a soil sample is measured with a known water content, compacted with a standard number of blows in a standard mould. The test is repeated for several moisture contents. By plotting dry density against moisture content (as a percentage of dry weight) the optimum moisture content and optimum dry density are obtained. If is often specified that the soil used as fill material shall be compacted to at least 95% of the optimum (Proctor) density. See SUPERFICIAL COMPACTION.

profile Usually a longitudinal section showing ground undulations, near-surface deposits, and possibly the planned formation level of a proposed construction, such as a canal or pipeline. See STREAM PROFILE, STREAM SECTIONS.

profile of equilibrium A stream with a profile along which all irregularities (waterfalls, etc.) have been removed by degradation, and all basins filled with sediment. See DRAINAGE EQUILIBRIUM, GRADED STREAM, REGIME.

179

profile, stream See STREAM PROFILE.

prograding The gradual building out to sea of a coastline by deposition of detritus; often the process is a temporary phase and not long continued. See COAST OF EMERGENCE, RETROGRADING.

propeller The curved rotating blade of a centrifugal pump. See IMPELLER.

propeller type current meter A meter incorporating an impeller which is rotated on a horizontal axis by the water flow. See CURRENT METER.

prospecting The search for minerals or other occurrences of economic value. Where promising localities have been located, the presence of good water supplies is an important factor in their development. See GEOPHYSICAL PROSPECTING, WATER PROSPECT MAP.

protection forest A track of ground covered with trees and woody growth and retained primarily for its beneficial effects on soil and water conservation. See SPREADER STRIP.

provinces, groundwater See GROUNDWATER PROVINCES.

puddle (or **puddle clay**) A dense clay practically impervious to water and sometimes used to line earthen water channels and in the core walls of earthen dams and embankments; to make watertight by packing with CLAY PUDDLE. See HEELING, LINING.

puddle clay See PUDDLE.

puddling A process by which the structure of a soil is destroyed with a reduction in porosity and permeability. Sometimes used in canals, water supply channels, and reservoirs to reduce seepage losses. See REMOULDING.

pulsometer pump A steam-operated pump in which the only moving part is an automatic ballvalve. The valve admits steam alternately to a pair of chambers, so forcing out water which was sucked in by condensation of the steam after the previous stroke. It can deal with dirty water and was widely used for miscellaneous pumping where steam was available. In general, it has been superseded by rotary pumps. See SUBMERSIBLE PUMP.

pump A machine or mechanical appliance for imparting energy to water or for raising water from a lower to a higher level in a pipe. See CENTRIFUGAL PUMP, FORCE PUMP, LIFT PUMP, PULSOMETER PUMP, ROTARY PUMP, SINKING PUMP, TURBINE PUMP.

pumpage The rate at which water is withdrawn from a well, shaft, or other excavation by a pump; the volume of water removed by pumping during a specified period. See DRAWDOWN, ECONOMIC YIELD.

pumped storage Term applied to various storage schemes, such as: (1) A means of storing the energy produced in thermal power stations. Water is pumped up to a high-level reservoir during the night when surplus power is available from steam stations and released during the peak load period of the day through water turbines to generate power. (2) A form of storage sometimes used where the topography is not suitable for the construction of an impounding dam. The water is pumped from one or more rivers and stored in earthen reservoirs sited a short distance from the river. Sometimes a shallow valley in the area is utilised for storage with an earth dam and the water pumped into it. (3) Increasing the natural drainage to a reservoir by pumping water from an intake in an adjoining catchment area. The total yield is the pumped supply plus the runoff from the reservoir catchment. See CATCHWATERS.

pumped-storage hydro station   A hydroelectric plant in which surplus low-cost energy is stored for use in periods of peak demand. The layout consists mainly of a low-level and a high-level reservoir which are connected by water conduits (usually underground tunnels). During peak load periods, water is drawn from the high-level reservoir through the conduits to operate the generating plant, while during off-peak periods water is pumped back from the low storage to the high storage. The scheme has the advantage of requiring relatively small storage reservoirs. There are several pumped-storage hydro stations in Europe, and the U.K. has two outstanding examples at Cruachan in Scotland (completed in 1965) and at Ffestiniog in Wales (completed in 1962).

pumping (land drainage)   The use of pumps to drain low-lying areas or where the slope for gravity flow is inadequate. The pump may discharge its water into the nearest stream or on to land where gravity flow is possible. In major schemes, the pump may discharge into a central pumping station, which in turn delivers the water from several pumping units into a river, lake, or other storage.

pumping level (well)   The level from which water has to be pumped after the CONE OF DEPRESSION has been established in the local water-table.

pumping test   (1) Yield of water. A test made with a pump in a new well to determine its water-yielding capacity. Quantities and water levels are recorded during the test period. The test pumping rate is usually greater than that at which water will be required and covers a period sufficiently long to indicate whether the yield can be maintained. See ECONOMIC YIELD. (2) Quality of water. Taking water samples during the test to determine, by chemical analyses, the chief constituents and organic purity. Tests may extend over about 14 days and finally a full mineral analysis is often made and may be used to prescribe treatment and purification processes. See WATER QUALITY.

Q

quality (snow)   The amount of ice in snow, given as a percentage of the weight of the SNOW SAMPLE.

Quaternary (or **post-Pliocene**)   The most recent deposits in the geological column; consist of gravels, river alluvium, raised beaches, peat bogs, blown sands, morainic deposits, boulder-clays, glacial sands, etc. These materials are handled in most water engineering schemes and pre-knowledge of their nature and thickness is important during the planning stage. See DAM, ENGLACIAL MATERIAL, ICE AGE, RAISED BEACHES, TERRACES.

quick condition   Applied to a soil condition in which an upward flow of water has reduced its intergranular pressure and also its bearing capacity. See UPLIFT.

quick-return flow   Sometimes applied to that portion of rainfall which enters the upper soil layers and moves laterally into a stream. See SHALLOW PERCOLATION, SUBSURFACE RUNOFF.

quicksand   An unstable, waterlogged sand with no bearing capacity; a sand in which the upward movement of water is so strong that the mass is held in suspension by the water. The most susceptible sands in this respect are uniformly fine-grained with an effective size less than 0·1 mm and a COEFFICIENT OF UNIFORMITY below 5. Drainage will improve the stability of a quicksand. See BOIL, SAND.

quoin post   See HEEL POST.

R

race   A rapid or strong current of water or the conduit or channel for such a current; a stream from a pond or reservoir to a water wheel which it drives. See TAIL RACE.

rack   See TRASH RACK.

radial drainage   A natural drainage pattern associated with the updoming of a relatively level area with concentric belts of weak and relatively hard beds. The radial drainage was initiated on the domed surface, with streams flowing outwards in nearly every direction. In many cases, the simple dome structure has almost completely disappeared as the streams cut downwards and flow on older rocks. See DRAINAGE PATTERN.

radial gate (or segmental sluice gate or Tainter gate)   A gate located on the spillway crest of a dam with a curved upstream face which is horizontally pivoted. The curved surface is made concentric to the pivot or pin hence the resultant water pressure passes through the pivot. The design simplifies the hoisting of the gate (Figure R.1).

radial well   A well where a number of strainer pipes are driven laterally into a water-bearing deposit in radial fashion from a main sump. See STRAINER.

radius, hydraulic   See HYDRAULIC RADIUS.

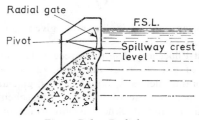

Figure R.1.   Radial gate

radius of influence (water well)   The radius of the circular base of the CONE OF DEPRESSION formed in the groundwater around a well when pumping is in progress.

raft   A flat floating structure, consisting of logs, casks, etc., for conveying persons and materials across a body of water; used in emergencies as a substitute for a boat.

raft bridge   A bridge in which rafts are used as supports.

raft foundation   A type of foundation often used under heavy structures or in soft, wet ground. It consists of a continuous slab of concrete, generally reinforced, extending under the entire structure. See BUOYANT FOUNDATION.

rain   Condensed moisture of atmosphere falling as liquid droplets.

rain day   A period of 24 hours during which a rainfall rate which is greater than a specified amount is recorded; term used only in some countries.

raindrop erosion   See SPLASH EROSION.

rainers (or rain guns)   Large sprinklers with a high output of water and a high working pressure, giving a radius of circle watered from 18 to 55 m (60 to 180 ft). Since only large water drops will reach the outer rim of the circle watered, care is required to prevent soil damage. See SPRINKLER IRRIGATION.

rainfall   Liquid droplets falling from the atmosphere or clouds during a rainstorm; the volume of rainfall over a specified period. Statistics show that for the 35 years 1916–50 the average rainfall for Great Britain is 1064 mm (41·9 in) per year. See CONVECTIONAL RAIN, DRIZZLE, OROGRAPHIC PRECIPITATION, PRECIPITATION, STORMWATER.

rainfall area   The storm area.

rainfall, average   See AVERAGE ANNUAL RAINFALL.

rainfall distribution coefficient   A storm coefficient obtained by dividing the maximum rainfall at any point within the storm area by the average rainfall in the area.

rainfall, duration   See DURATION OF RAINFALL.

rainfall excess   The DIRECT RUNOFF or the amount of rainfall exceeding that taken by infiltration, evaporation, and absorption. Rainfall excess investigations have been aided in recent years by the use of high-speed computers and also by rainfall runoff models. See RATE OF RAINFALL, SURFACE RUNOFF.

rainfall index   Index representing the combined effects of DEPRESSION STORAGE, INFILTRATION and INTERCEPTION; average rainfall above which the volume of rainfall equals the observed runoff.

rainfall intensity   See RATE OF RAINFALL.

rainfall intensity curve   See INTENSITY CURVE (rainfall).

rainfall intensity, maximum   See MAXIMUM RAINFALL INTENSITY.

rainfall penetration   The depth to which a given quantity of rainfall infiltrates. The maximum possible penetration is down to the water-table. See INFILTRATION RATE.

rainfall province A region or district throughout which the seasonal concentration and distribution of rainfall are similar.

rainfall rate See RATE OF RAINFALL.

rainfall recurrence interval See INTENSITY FREQUENCY OF RAINFALL.

rainfall, seasonal variations The average rainfall for any locality varies from month to month. The statistics of the monthly falls at Greenwich (U.K.) averaged over a period of 100 years (1820–1919) indicate that October is the wettest and February and March the driest months. The evaporation loss of rainfall during the summer months is high. Again, average rainfall at a mountain station is considerably higher than at valley stations. See ISOHYETALS.

rain, freezing See FREEZING RAIN.

rain-gauge (or **gauge** or **pluviometer** or **udometer**) A device which collects and measures the amount of precipitation and expresses it as depth of rain to hundredths of an inch or tenths of a millimetre, and includes the equivalent depth of melted snow or hail. It may consist of a funnel from which the water drips into a graduated cylinder. In the U.K. rain-gauges are usually 127 mm (5 in) in diameter and set with their rims 305 mm (12 in) above ground level and placed at sites which are not too exposed nor too sheltered. See BRADFORD GAUGE, PLUVIOGRAPH, SHIELD.

rain-guns See RAINERS.

rainmaking See CLOUD SEEDING.

rain recorder An instrument which gives a continuous record of precipitation. The trapped water elevates a small float to which a pen is attached for tracing a continuous record on a clock-driven chart. Also called a UDOMOGRAPH. See GAUGE.

rain, residual See RESIDUAL RAIN.

rainshadow A meteorologically dry area sometimes induced on the leeward side of a line or range of high country. See NON-FRONTAL PRECIPITATION.

rain, trace See PRECIPITATION, TRACE OF.

rain-wash (1) The action of rain in loosening and washing down rock and soil material to lower levels and finally into reservoirs, rivers, or lakes. (2) A landslide of broken rock and soil caused by heavy rainfall. See GULLY DRAINAGE, SHEET EROSION, SOIL EROSION.

raised beaches Ancient beaches which now stand well above sea level owing to slow relative uplift of the land. Around the Scottish coasts, raised beach terraces occur at elevations ranging up to 30 m (100 ft) above the present sea level, while in Norway and Sweden some beaches lie at over 180 m (600 ft) above the present shore. See TERRACES.

raising of dams See DAMS, RAISING OF.

ram (or **plunger**) The moving cylindrical block in the working chamber of a RAM PUMP or FORCE PUMP.

ramp A short length of drain with a much steeper slope than normal.

ram pump See FORCE PUMP.

range Used mainly in the U.S.A. for a stretch of grazing land with little subdivisional fencing; cover composed mainly of herbs, native grasses, or shrubs. See GRAZING MANAGEMENT.

range of tide, great diurnal See GREAT DIURNAL RANGE OF TIDE.

rapid flow (or shooting flow) Flow in open channel with a velocity greater than that of a gravity wave; flow in open channel in which FROUDE NUMBER is greater than unity.

rapids The swift and sudden descent of a stream without actual waterfall; may be formed when a stream passes over a resistant rock with softer beds below it; the greater erosion of the latter develops sufficient drop in slope or grade to produce rapids. See CASCADE, NARROWS, WATERFALLS.

rate of flow See DISCHARGE.

rate of infiltration See INFILTRATION RATE.

rate of rainfall (or intensity of rainfall or rainfall intensity) The rate of rainfall expressed as depth of water in inches or millimetres per hour, day, or other specified period. See MAXIMUM RAINFALL INTENSITY.

rate of runoff The rate at which water flows from a catchment area at any given time; usually in cumecs or cusecs. See PEAK RATE OF RUNOFF.

rating (1) In general, the relationship between two mutually dependent quantities. In hydraulics, e.g., the relationship between the water level and the discharge of stream or channel, or between CURRENT METER revolutions and water velocity. (2) Taking flow measurements to establish a rating. See CURRENT METER RATING, DURATION CURVE.

rating curve A curve or tabular representation of RATING. See DURATION CURVE.

rating flume A CONTROL FLUME; a flume containing still water along which CURRENT METERS are drawn at known velocities in order to calibrate them; an open conduit constructed in a channel to ensure good control for flow measurements and values for a RATING CURVE.

rating loop A DURATION CURVE which describes a loop with each rise and fall of the river; caused by higher discharge values for given water levels when river is rising than when it is falling.

rating table A table showing the relationship between two mutually dependent variables or values, e.g., between discharge of a stream and water level at a gauging station. See RATING.

rating tank A tank containing still water through which a CURRENT METER is drawn at a known velocity in order to calibrate the meter.

rational runoff formula A U.S.A. formula to calculate stream discharges from small catchments:

$$Q = AIC$$

where Q is the peak discharge in cusecs, A is the area of catchment in acres, I is the average intensity of rainfall in in/h during the period required for all parts of the catchment to contribute to runoff, and C is the runoff coefficient usually considered as that part of the rainfall rate which will contribute to the runoff rate. See LLOYD-DAVIES FORMULA.

raw bank A river or other bank in its natural state or where the protective surface has been eroded or stripped off.

raw-bank system A heavy timber construction to provide protection against bank erosion. Tree trunks, placed with butts upstream, are secured by steel cable to vertical bearer logs along the bank and also to posts set well back from the bank. The bearer logs are trenched and layered down-bank with toes below low water level. Individual trees are wired together with not less than two ties per tree. Used in Australia and New Zealand. See ANCHORED TREES, BATTERED-BANK SYSTEM.

reach The distance between two marked or specified stations or cross sections of a river or stream; the unobstructed stretch of water between two locks. See POND.

reaction turbine A WATER TURBINE in which the nozzles or jets are on the moving part and not fixed. See IMPULSE TURBINE.

realignment A change in the planned direction of a tunnel, canal, pipeline, or other hydraulic excavation. A change in direction (horizontal alignment) often affects its gradient (vertical alignment).

recession curve, normal See HYDROGRAPH, RECESSION.

recession hydrograph See HYDROGRAPH, RECESSION.

recession of coast The retreat of the coast as a result of wave erosion; may be very rapid along sandy and low-lying areas. Protective work includes different forms of walls, jetties, and bulkheads with varying results. See COAST OF SUBMERGENCE.

recharge of aquifer (or groundwater increment or groundwater recharge) The introduction of surface water, such as streams and rainfall, into underground storage. Inflow of surface waters must be geologically related to the aquifer to be recharged. This requires knowledge of the local geology, structures, and the trend of the permeable and impermeable deposits.

recharge of aquifer, induced (or induced recharge of aquifer) Flow of stream water into aquifer.

recharge of basin See BASIN RECHARGE.

recharge of groundwater See GROUNDWATER INCREMENT, SEASONAL RECOVERY.

recharge well An INVERTED WELL which conducts surface water into an aquifer at shallow or moderate depth; a type of well often used in arid regions to conserve water. See INVERTED CAPACITY.

reciprocating pump Any type of pump which lifts water to a higher level by a series of to-and-fro movements of a piston or plunger. See PUMP.

recorder See RECORDING GAUGE.

recording gauge (or recorder) (1) An instrument which automatically records values, quantities, or conditions, such as rainfall, flow, stage, pressure, or temperature. The values are usually given as a graphic representation on a time basis. (2) A water-level recorder shows the level of water in any conduit, reservoir, or tank over a specific period (usually 7 to 28 days); often operated by a float and levels shown on a chart. See GAUGE, WATER STAGE RECORDER.

recovery, cyclic See CYCLIC RECOVERY.

recovery, seasonal See SEASONAL RECOVERY.

rectangular crib groyne A CRIB GROYNE, rectangular in shape, and filled with stone. The length usually exceeds 6 m (20 ft) and the width between the side walls is about 1 to $1\frac{1}{2}$ times the height. They can be made with continuous sides and wire or cross-logs used to prevent spreading. Construction is similar to triangular groynes and wire, rods, or spikes are used to bind the logs and walls together. The bank is sloped and grassed. See GROYNES, TRIANGULAR CRIB GROYNE.

rectangular drainage A natural drainage pattern where the course of the streams is controlled by fault and joint systems in the superficial deposits, or by folded or tilted strata consisting of alternating strong and weak beds. See DRAINAGE PATTERN.

rectangular weir A MEASURING WEIR in which a thin plate of rectangular shape is used extending perpendicular to the direction of flow.

recurrence interval The average number of years within which a given hydrological event, such as flood, is equalled or exceeded. See N-YEAR FLOOD.

recurrence interval (rainfall) See INTENSITY FREQUENCY OF RAINFALL.

reed A reed *(Phragmites communis)* common in the U.K. and consisting of tall, stiff stems; cut and tied into bundles of about 0·75 m (30 in) girth and 2·1 m (7 ft) long to form a DUTCH MATTRESS. See FASCINES.

reefing float See CAPTIVE FLOAT.

reference current meter During the measurement of discharge of a stream the current meter is immersed in a fixed position. During the gauging operation, it is assumed that the change in velocity shown by the reference current meter is proportional to the change of discharge when the latter is small.

refilling See BACK FILLING.

reflux valve See CLACK VALVE.

regelation (of ice) The melting of ice by an increase of pressure and, secondly, its solidification or regelation when the pressure is reduced or removed.

regenerated water (or return water) Water diverted from a stream or other surface source, for irrigation or other use, which seeps downwards and laterally to return to a stream or other body of water.

regeneration curve One showing the rise in water level in a well, plotted against time, after pumping has ceased. See CONE OF DEPRESSION.

regime (1) Applied to a canal or stream in which the water flow is such that it neither deposits material on its bed nor picks up material from it. At any particular period, only short lengths of a stream are in regime, while canals after some years tend towards the condition. See PROFILE OF

EQUILIBRIUM. (2) Prevailing character of climate or of some climatologic factor, such as seasonal distribution of rainfall. Also called REGIMEN (U.S.A.).

regime, coefficient of See COEFFICIENT OF REGIME.

regimen See REGIME.

regular velocity distribution The velocity distribution between two defined cross sections of a channel which does not change for a given flow.

regulated flow A condition where upstream reservoirs regulate the flow in a stream. See BASE FLOW.

regulation (stream) Where the natural flow in an open conduit or stream is controlled.

regulator storage See PONDAGE.

reinforced concrete CONCRETE in which metal (usually steel rods) is embedded such that the two materials act together in resisting loads or forces. Concrete alone is strong in compression but very weak in tension. Stress calculations are usually made on the assumption that the concrete takes all the compressive forces and the steel takes all the tensile forces. See PORTLAND CEMENT.

reinforcing fabric Usually steel mesh, which may be interlaced, woven, or welded, and used to secure materials such as vegetation, stone, etc. for river bank protection or for securing and reinforcing certain types of gornyes and other works. See GROYNES, SAUSAGE CONSTRUCTION, STONE-MESH GROYNES.

rejuvenated water A term sometimes applied to WATER OF COMPACTION and water released during metamorphism. See MAGMATIC WATER.

rejuvenation Applied to a stream where the velocity of flow has increased, as by uplift or tilting of the land, with a renewal of downward erosion of its bed. It often cuts below the level of the plain, parts of which are left as river terraces.

relative humidity The ratio, expressed as a percentage, of the water vapour in the air to that in saturated air. Relative humidity is the most convenient method of expressing humidity for purposes of water supply. See HUMIDITY.

relief channel (rivers) A passage, canal, or conduit cut between favourable points along the course of a river and used to store water in times of flood and discharged when the level or tide was low. See BARRAGE.

relief well A borehole put down at the toe of an earth dam to lower any high PORE WATER PRESSURE formed by the weight of the dam (Figure R.2).

remoulding The derangement of the internal structure of a silt or clay, with the result that it gains compressibility but loses shearing strength. See PUDDLING.

removal of support The removal of ore or coal by mining operations under a reservoir or canal. Within a certain depth, such removal of support may cause severe damage to the surface works and may result in a costly legal action.

188

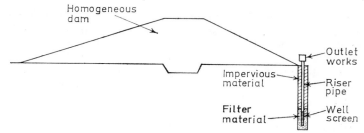

Figure R.2. Relief well

remove and stockpile topsoil An earth dam contract term which is only used to indicate topsoil that is required for re-use as cover.

required excavation A contract term in dam construction which includes all excavation for the foundations and for structures, but excludes quarries, borrow areas, and temporary works.

reservoir An artificial lake into which water drains and is stored for future use; a tank, basin, cistern, mill pond, or any place in which water has accumulated. A reservoir provides for the storage of irregular sources of water, such as flood and storm water and intermittent streams, and also provides storage capacity during non-irrigation periods. See DAM, FARM POND, FARM WATER SUPPLY ENGINEERING, FAULT, IMPOUNDING RESERVOIR, SERVICE RESERVOIR, SILTING, SURFACE WATER SUPPLIES.

reservoir basin The area of low-lying land or valley to be flooded at a dam site. Detailed geological maps are prepared to indicate any faults, zones of weakness, or pervious rocks. The permeability of the rocks may be determined by tests, as by injecting or pumping water down boreholes under pressure and observing the actual losses. See DAM SITE INVESTIGATIONS.

reservoir capacity The volume of water that can be stored in a reservoir up to FULL SUPPLY LEVEL, and usually expressed in cubic metres, acre feet, or millions of gallons. A low stream gradient gives a much larger capacity than a steep gradient for any given height of dam. See STORAGE CAPACITY.

reservoir demand The amount of reservoir water required from time to time by the consumers. Ideally, the RESERVOIR DRAFT should be equal to the reservoir demand.

reservoir draft The amount of water withdrawn or released from a reservoir periodically to meet the requirements of consumers. See STORAGE PERIOD.

reservoir inflow The water which flows naturally or artificially (e.g. by pumping) into a reservoir over a given period.

reservoir inflow hydrograph A graph showing RESERVOIR INFLOW over a given time period.

reservoir, life See LIFE OF RESERVOIR.

reservoir losses Water held in storage is subject to losses, mainly by seepage through the reservoir floor and embankments and also by evaporation from the water surface. See PERCOLATION PATH (DAMS).

reservoir sedimentation   The deposition of silt and sediment transported by streams, runoff, etc., into a reservoir. See SILTING, LIFE OF RESERVOIR.

Reservoirs Safety Provisions Act 1930 This U.K. Act requires that a reservoir designed to hold or capable of holding 5 million gal (22 500 m³) or more of water above the natural level of any part of the adjoining land must be designed and constructed under the supervision of, and inspected from time to time by, a qualified civil engineer on one of the panels constituted under the Act. See LARGE DAM.

residual mass curve   A curve showing year-to-year residual departures of discharge or rainfall from arithmetic average accumulated for a given period. See MASS DIAGRAM.

residual rain   Rainfall near the end of a storm at a rate less than INFILTRATION CAPACITY. See INITIAL RAIN, NET SUPPLY INTERVAL.

resistance   The friction which flowing water must overcome because of the roughness of the channel or pipe; expressed in metres (feet) head of water. See CHANNEL ROUGHNESS, ROUGHNESS COEFFICIENT.

resistance strain gauge   See ELECTRICAL RESISTANCE STRAIN GAUGE.

resistivity method   A GEOPHYSICAL PROSPECTING method in which direct measurements are made of the ratio of voltage to current when a current is forced to flow through the ground to be tested. The conducting property of a rock is controlled by its water content and its salinity. If these values are high then its conductivity is also high and its electrical resistivity is low. The method has been used extensively for water and foundation problems. In certain conditions the SEISMIC METHOD has received preference.

resoiling   The replacement of the original topsoil or of fresh soil along the bed and banks of rivers, canals, and reservoirs on completion of operations to allow the growth of vegetation to reduce erosion and increase stability of slopes.

response time (storm)   See PERIOD OF CONCENTRATION.

rest water level   Sometimes applied to the STANDING WATER LEVEL.

resurgent water   Sometimes applied to MAGMATIC WATER.

retained water   Water retained in a soil or rock by molecular attraction against the force of gravity; the water remaining in a soil or rock after the GRAVITY GROUNDWATER has drained out.

retard   A permeable spur built out from the bank into a channel to deflect the main flow of water away from the bank to reduce erosion. See JETTY, REVETMENT, TREE RETARDS.

retarding reservoir (or **retarding basin**)   A reservoir with uncontrolled outlets, for the temporary storage of flood water. See DETENTION RESERVOIR.

retention   The difference between total precipitation and total runoff on a drainage area. See INFILTRATION, INFILTRATION COEFFICIENT.

retention reservoir   A reservoir in which water may be stored or released as desired by gated outlets below spillway level.

retention, specific   See SPECIFIC RETENTION.

retention, surface   See INITIAL DETENTION.

retrograding Active erosion on a coastline where the shore line is cut back into the land mass. See PROGRADING.

retrogression (1) The lowering of bed levels of streams by downward erosion. (2) The lowering of water level at the same discharge. See AGGRADATION.

retrogression, sympathetic See SYMPATHETIC RETROGRESSION.

return period A measure of the most probable time interval between occurrence of a flood of given magnitude and that of an equal or greater flood. See EXCEEDANCE INTERVAL, FLOOD (RETURN PERIOD), FLOOD PROBABILITY.

return water See REGENERATED WATER.

re-use of water The use of the same water several times. The purification of industrial water is performed by chemical treatment or filtration or other methods. The water obtained from rivers is often used several times before it re-enters the river.

reversed filter See LOADED FILTER.

reverse flow A flow in channel or conduit opposite to normal flow.

revet To reduce or prevent the wearing away of a surface or bank with a REVETMENT.

revetment A lining or wall built along the sides of a stream to protect the bank and minimise erosion; may be temporary until the bank can be stabilised by solid works or live vegetation. Also applied to any covering to prevent scour of rock or soil surfaces such as concrete slabs, grass, maritime plants, faggots, asphalt sheets, etc. See APRON, FASCINES, PITCHING, RETARD, RIPRAP, STONE RIPRAP.

Reynolds number (or Reynolds criterion) A dimensionless number symbolised by (Re), or, less preferably, R, named in honour of Professor Osborn Reynolds and applied to fluid flow in a tube. The value (Re) is the same for all fluids at the CRITICAL VELOCITY.

$$Re = \frac{\text{mean velocity of flow} \times \text{pipe diameter}}{\text{kinematic viscosity}}$$

The number is sometimes used in MODEL ANALYSIS, and should be the same in both model and corresponding structure. See FROUDE NUMBER.

rias See DROWNED VALLEYS.

ria type coast A coastal area with long parallel estuaries (or rias) running far inland, as in south-western Ireland; developed where the river valleys and ridges are at right angles to the coastline. See DALMATIAN TYPE COAST, FIORD COAST.

ridge terrace An embankment of earth to reduce erosion by diverting runoff water across the slope of the ground instead of allowing it to flow directly down the slope. It usually consists of a long, low ridge of earth with a gentle slope along the sides, and a shallow channel along the crest. See BENCH TERRACE, BROAD-BASE TERRACE, NICHOLS TERRACE.

ridging for irrigation Controlling the flow of irrigation water in fields by constructing small ridges, borders, or embankments.

right bank The right-hand bank of a river facing downstream (Figure A.4).

rigid joint A joint in a pipe which allows no movement. The type includes welded, flanged, and turned-and-bored. See SEMI-RIGID JOINT.

rill erosion Soil erosion by rills or streamlets; the rills can be levelled out by normal cultivation. See GULLY EROSION.

ring dam An earthen dam, usually of circular shape, built on level ground from material excavated from within the storage basin. to impound water for farm supplies and irrigation. These small dams may have STORAGE RATIOS varying between 2 : 1 and 6 : 1 (Figure R.3). See HILLSIDE DAM, TURKEY'S NEST DAM.

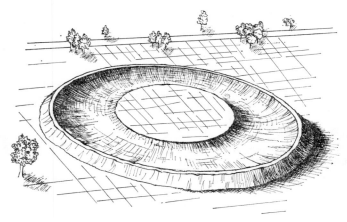

Figure R.3. Ring dam

ring levee An artificial, irregular ring-shaped LEVEE sometimes used to protect an area of land enclosed by a river and an effluent which later rejoins the main river. The ring levee is built to protect the island so formed.

ring main A main irrigation pipe installed in the form of a complete circle. It enables a greater area to be irrigated than in the case of a single line. Friction losses are reduced, since the water has two conduits to reach each distribution point. The pipe column can thus be of smaller diameter, although its total length is greater. A ring main may also be used in a water supply system. A complete circuit is the aim, in which case a supply is available in either or both directions. It is often connected to all or most of the TRUNK MAINS in the area.

ring tank See RING DAM.

riparian Situated on, or pertaining to, the banks of a river or body of water; the rights or ownerships along the banks of rivers, lakes or canals; a dweller on a river bank. A riparian owner has rights over part at least of one bank.

riparian land Usually the land situated along the banks of a natural body of water.

riparian rights The rights of a riparian owner to the use of a stream or other water which borders or flows through his land. See WATER RIGHTS.

riprap (or enrockment) A covering or backing of broken stone or brush and stone on the face of a dam or the bed or banks of a river as a protection against erosion; often used along irrigation channels or river improvement works. See REVETMENT, STONE RIPRAP.

rising tide See FLOOD TIDE.

river A stream of water, larger than a brook or rivulet, flowing in a natural open channel towards the ocean, a lake, or another river. In addition to the part it plays in draining land areas, a river may supply large quantities of water for industry and towns. From a legal or administrative point, the term 'river' may include any lake or artificial storage along the river. See ALLUVIAL RIVER, BRAIDED COURSE, CONSEQUENT STREAMS, CREEK, INCISED RIVER, OLD RIVER, SUBSEQUENT STREAM, TIDAL RIVER.

river bank erosion See EROSION, STREAM BANK EROSION.

river basin The land area which a river and its tributaries drain. See CATCHMENT AREA.

river bed That relatively flat part of a river channel where water normally flows.

river breathing See BREATHING (of river).

river capture Applied to a river which cuts into the headwater of another, less active, river and captures some of its tributaries. River capture or piracy occurs because of unequal conditions, such as nature of rocks, difference in slope or size of rivers. See 'MISFIT' STREAM.

river channel See BANK, CHANNEL, RIVER BED.

river control See RIVER ENGINEERING, RIVER IMPROVEMENT, RIVER TRAINING.

river diversion tunnel A tunnel to divert the river during the construction of a dam. It is excavated into the hillside and around one end of the new dam and has a capacity to take flood flows. The upstream or inlet end is covered with a concrete or metal grid to stop debris and trash.

river engineering The branch of civil engineering which deals with the control of rivers and their improvement and flood mitigation. A full appreciation of a river's characteristics demands a knowledge of local geology and hydrogeology. See BENEFIT RATIO, RIVER IMPROVEMENT, LAND ACCRETION, RIVER TRAINING.

river flats (or flats) Deposits of silt, gravel, etc. bordering a river along its lower reaches and sometimes subject to flooding to a greater or lesser degree. See ALLUVIAL FLAT, DELTA-PLAIN SWAMPS.

river floodplain swamps See DELTA-PLAIN SWAMPS.

river forecasting Estimating or forecasting river discharge and water levels, based on HYDROLOGY and METEOROLOGY. See HYDROMETEOROLOGY, RIVER ENGINEERING.

river gauge See GAUGE.

river gauging See GAUGING, STREAM GAUGING.

river improvement Mainly, engineering works and practices to improve the flow conditions of a river and to control erosion and siltation. The works may include deepening, widening, straightening, or diverting the channel; prevention or reduction of damage by flooding; control of erosion of banks and bed; removing obstructions from bed and banks to improve flow; construction of groynes, weirs, levees, revetments, etc. for river training; construction of new outlets for river, canalisation, etc. and also the planting of trees, shrubs, or grasses for protection of bed, banks, or associated works. See RIVER ENGINEERING.

river loads See LOAD.

river piracy See RIVER CAPTURE.

river profile See STREAM PROFILE.

river stage The height of the water surface relative to a fixed datum. See HOOK GAUGE, STAGE.

river terraces Natural benches or terraces which border many rivers; usually narrow but often of considerable length along the river. Several terraces may be present on one or on both sides and they usually represent the remnants of FLOODPLAINS.

river training Mainly concerned with engineering works to improve the tidal and non-tidal sections of rivers for navigation and other purposes. Improvements may include dredging to remove shoals and to secure deeper water; restricting the flow to a selected channel by construction of training walls through an estuary, or by the use of low groynes to concentrate the flow in the main stream. In general, efforts are made to concentrate the scouring effect of the current and to keep the channel clear and deep. See FLOOD CONTROL.

rivulet A creek, brook, or small stream.

roaded catchment (Figure R.4) See ARTIFICIAL CATCHMENT.

roadside erosion control See HIGHWAY EROSION CONTROL.

rock and soil symbols See MAP SYMBOLS.

rock excavation An earth dam contract term which includes all solid bedrock which has to be loosened by blasting. It cannot be broken up by heavy pneumatic equipment and includes boulders over $0·765$ m^3 (1 yd^3).

rockfill dam Defined as a fill dam composed of more than 50% of pervious rockfill material. The type has many modifications. The watertightness is provided by an impervious membrane or facing which is usually placed on the upstream face of the dam and is connected to a CUTOFF WALL.

rockhead Another name for BEDROCK.

194

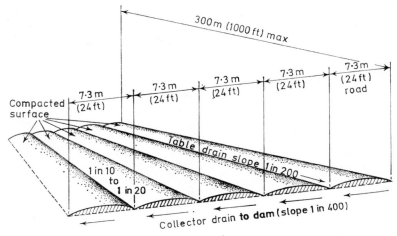

Figure R.4.   Roaded catchment

rodding   Cleaning out drains with DRAIN RODS which are inserted through a manhole or cleaning eye. See INSPECTION CHAMBER.

rod float   See FLOAT ROD.

roller   A heavy vehicle used for compacting soil and earth dams to increase density and stability of the material. See SUPERFICIAL COMPACTION.

roller gate   A CREST GATE, in the form of a hollow cylinder, for regulating the flow at dam spillways; carried at each side on large toothed wheels which mesh with steeply inclined racks up which the gate moves when being opened. See SECTOR GATE, SLIDING GATE.

rolling up curtain weir   A type of FRAMED DAM in which the frame remains upright. It is built up with horizontal planks, connected by chains, and drawn up when being opened. The planks increase in thickness with their depth below water.

rollway An overflow spillway or the overflow portion of a dam.

root   The part of a dam construction which becomes rooted or embedded in the ground where it joins the hillside. See DAM FOUNDATION.

root concentration   An expression of the root concentration in a soil. The soil is usually divided into 150 or 300 mm (6 or 12 in) thick zones and the percentage of the total root system given for each zone.

root of groyne   The end of the groyne buried or rooted in the bank of a river. See GROYNES.

root zone   That soil horizon or zone in which 90% of plant roots are located.

roster system   A system in which irrigators receive their supply of water on fixed dates during the season. The dates are shown on cards supplied to the irrigators. See ROTATION IRRIGATION.

195

rotary drill A fast and continuous method of drilling a deep well. The drill bit remains in the hole and additional drill stem is added as the hole deepens. A mixture of clay and water is forced through the bit. The cuttings are carried upward to the surface, and into a tank where they settle out, and the mud is pumped back into the well. When the drilling is complete, a perforated casing is inserted in unconsolidated aquifers to support the walls and allow water to enter the well. The drill is operated by electricity or compressed air. See CABLE TOOL.

rotary meter An instrument for measuring the velocity of flow in a river or any open channel. See CURRENT METER.

rotary pump A pump consisting of rotating elements and capable of raising large volumes of water at low pressure. It does not depend on centrifugal force to lift the water and it may operate at almost any speed. See PUMP, TURBINE PUMP.

rotational slide See FELLENIUS'S CIRCULAR ARC METHOD.

rotation irrigation A scheme by which a number of irrigators, receiving water from the same lateral, agree to take the entire flow in turn for a limited period. See CONTINUOUS DELIVERY.

roughness coefficient (or coefficient of rugosity) A value indicating the channel roughness; used in certain formulas, such as BAZIN'S, KUTTER'S and MANNING'S, when calculating the velocity or flow of water in a stream or channel.

roughness of channel See CHANNEL ROUGHNESS.

round-head buttress dam A form of concrete BUTTRESS DAM consisting of a radial upstream water-supporting face which transmits the water pressure to a buttress below. See MULTIPLE-ARCH-TYPE DAM.

routing flood See FLOOD ROUTING.

rubble Any mixture of concrete or brick fragments or small stones of irregular size and shape.

rubble concrete Concrete in which large blocks of stone are placed with a minimum of 150 mm (6 in) of space between them so that the concrete may be properly rammed and leaving no voids. The stones are roughly squared and arranged in courses, so that they break joint. This form of concrete is often used for the hearting of masonry dams.

rubble drain (or blind drain or spall drain or stone drain) A drainage trench or channel filled with rubble or stones of suitable size through which the water flows.

rubble-mound breakwater (or mound breakwater) A common type of BREAKWATER composed of stones weighing up to 5 t deposited at a nearly flat slope between tide marks and at a steeper slope below low water. The under water part is often stabilised by injecting fluid bitumen and sand at 200 °C. See BLOCKWORK.

rugosity See CHANNEL ROUGHNESS, ROUGHNESS COEFFICIENT.

rummel See DRY WELL.

runner The rotating part of a WATER TURBINE or other hydraulic wheel.

runoff That part of precipitation which flows from a catchment area and finds its way into streams, lakes, etc. It includes both DIRECT RUNOFF and GROUNDWATER RUNOFF. The amount of runoff is affected by (1) condition and nature of soil or rock; (2) intensity of precipitation; (3) slope; (4) vegetation and (5) wind. A high runoff may be expected if the rainfall is sudden and the ground is saturated or frozen. By gauging the streams that drain a catchment area its runoff may be calculated. See CATCHMENT YIELD, RATE OF RUNOFF.

runoff coefficient See IMPERMEABILITY FACTOR.

runoff cycle That section of the HYDROLOGIC CYCLE between precipitation or snow melt over land areas and direct return to atmosphere through EVAPO-TRANSPIRATION or subsequent discharge through streams.

runoff, depth See DEPTH OF RUNOFF.

runoff, overland See SURFACE RUNOFF.

runoff percentage The total amount of water which flows from a catchment area given as a percentage of total rainfall on the area.

runoff plots Small and relatively normal plots of land used for measuring DIRECT RUNOFF and perhaps other determinations, such as soil erosion and removal of soluble materials.

S

saddlebag groynes GROYNES consisting of shingle or other material covered over with wiremesh which has ample overlap at the outer end and sides. Rounded stones are placed over the mesh which is then turned back and fastened. The mesh is thus weighed down by a sausage along each side and a mattress at the end. It has been used with some success in Victoria, Australia, where machines are often used for piling the shingle. See MATTRESS, SAUSAGE CONSTRUCTION, STONEMESH MATTRESS.

safe velocity See PERMISSIBLE VELOCITY.

safe yield (aquifer) See ECONOMIC YIELD.

sag pipe See INVERTED SIPHON.

salinity (groundwater) The content of totally dissolved solids (T.D.S.) in the water; measured by the electrical conductivity method. For sea water the content is about $3\frac{1}{2}\%$ or 35 000 parts per million or mg/l. In some areas salinity is a hazard, as it might rise to levels higher than could be tolerated for irrigation. See BRACKISH, pH-VALUE.

salmon ladder A provision made in some sluices, weirs, or other barriers to permit the free passage of fish. It consists of a series of small pools, arranged in steps, down which a constant stream of water flows. See FISHWAY.

saltation (streams) See LOAD.

salt balance The difference between the total solids removed by drainage water from land annually and the total dissolved solids brought in by irrigation water annually. See CRITICAL CONCENTRATION.

197

salt index A value or formula for determining the suitability of a water supply for irrigation.

salting The accumulation of excess salt in the soil with harmful effect on crops and vegetation; may be caused by the rise of saline groundwater to near-surface level. The soil may be cured by drainage and then LEACHING.

salt load The load of salt carried by a stream; usually expressed in tonnes or tons. See SALINITY.

salt slug The introduction of a solution of highly concentrated salt into a stream of relatively low or no salinity.

salt velocity method (or **Allen's method**) A method to determine water velocity by introducing concentrated salt brine at a station and recording the time it takes to travel a known distance downstream. The passage of the brine at the end points of the known distance is detected by the change in electric conductivity of the flowing water. See CHEMICAL GAUGING.

sampling station The site or cross section selected on a stream or open channel at which the dilution or cloud is measured, sampled, or observed. See DILUTION METHODS, GAUGING STATION.

sand A loose accumulation of particles, mainly quartz, with a grain size between 0·02 and 2·0 mm. Sands and sandstones are important deposits as water reservoirs and they allow the migration of water downwards and laterally. See MICROPORES, POROSITY, QUICKSAND, SILT.

sand blanket A covering layer of sand, about 150 mm (6 in) thick, to prevent the cracking of bentonite or clay in reservoirs, canals, and water channels when they are empty.

sand boil A BOIL may develop downstream of an earth dam or embankment with sandy foundations. A temporary remedy is to surround the boil with a wall of sandbags to develop a back pressure and so decrease the hydraulic gradient and stop the loss of sand.

sand catcher (or **sand-grain meter**) An instrument through which the water flows and any sand settles out and its quantity is measured. See SAND TRAP.

sand drain See VERTICAL SAND DRAIN.

sand filter A filter for purifying domestic water. It has a layer of coarse stone at the bottom, grading upwards to fine quartz grains at the top. The water flows slowly downwards and the upper surface is renewed periodically to remove bacteria. A similar type of filter is used for treating sewage effluent but the sand is coarser and air has access to the sand. See GRADED FILTER, SLOW SAND FILTER.

sand grain meter See SAND CATCHER.

sand pump See BAILER.

sand pump dredger See SUCTION DREDGER.

sand reef See BARRIER BEACH.

sand screen (boreholes) A tube type of screen for boreholes, generally non-ferrous, with narrow openings for the entry of water. It may be used with gravel packing. See PERFORATED-CASING WELL, TUBE WELL.

sand trap A basin-shaped enlargement in a stream, ditch, or channel, where the silt and sand carried by the water is deposited due to the drop in velocity; means may be included to remove such material. See DESILTING BASIN, DESILTING STRIP.

saturated air Air containing the maximum possible amount of water vapour at the given temperature; air at a relative humidity of 100%. See VAPOUR PRESSURE.

saturated soil A condition when all the interstices of a soil are filled with water.

saturated surface See WATER-TABLE.

saturation Applied to the condition when water (or other substance) has filled all the voids of a rock or soil and it can assimilate or contain no more. See HUMIDITY, ZONE OF SATURATION.

saturation deficit A measure of the 'dryness' of the air; expressed by:

$$\frac{100-RH}{100} p$$

where RH is the RELATIVE HUMIDITY and p is the SATURATION VAPOUR PRESSURE.

saturation gradient Slope of the SATURATION LINE.

saturation line (1) A line indicating the uppermost limit of flow of water through an earthen dam, ridge, and subsoil and marked on a cross-section of the structure or mass. (2) The WATER-TABLE.

saturation vapour pressure The pressure exerted by a vapour in a saturated atmosphere.

saturation zone See ZONE OF SATURATION.

saucer A large flat float tied to a ship to elevate and float it past a shallow area. See CAMEL.

sausage construction A type of STONEMESH CONSTRUCTION used for river bank protection. It consists of tubes of wire mesh filled with stone, preferably rounded. The tubes are laid parallel down the bank with the lower ends closed and then filled from the top end. The bank is usually battered down to a regular slope before the tubes are filled. Where possible the lower ends of the tubes are extended about 0·9 m (3 ft) beyond the foot of the bank across the bed of the stream. This prevents erosion and undermining along the base of the bank. The structure is sometimes reinforced by a growth of roots and branches. See CRIBWORK, UNDERWATER APRON.

sausage dam A dam composed of rock fragments which have been wrapped into cylindrical bundles with wire and laid in a vertical or horizontal position to form a barrier across the stream or channel. See ROCKFILL DAM.

schlammdecke See ZOOGLOEAL LAYER.

scour The wearing away of the coast and sea bed, or the banks and bed of a river, or of terraces, diversion channels, or dams, mainly by waves and water action. See EROSION, SHORE EROSION.

scour, bed See BED SCOUR.

scouring sluice (or **scour pipe** or **washout valve**) A sluice, pipe, or opening, in the lower part of a dam, controlled by a gate, through which accumulated silt, sand, gravel, or other material may be occasionally ejected. It may also be used to lower the water level in the reservoir in an emergency.

scour pipe See SCOURING SLUICE.

scour protection Appliances and practices to prevent scour of rock or soil surfaces. See REVETMENT, VEGETATIVE PRACTICES.

scour valve A valve or outlet provided at low points in pipelines through which suspended solids or silt-carrying waters are being delivered.

scow See HOPPER BARGE.

scraper A steel tractor-driven surface vehicle, mounted on large rubber-tyred wheels, with a 4·6–9 m³ (6–12 yd³) capacity. The bottom incorporates a cutting blade which when lowered is dragged through the soil or earth. When full, the machine moves to the discharge point where the material is dumped through the bottom of the vehicle in an even layer. Used for stripping and re-levelling soil and earth dam construction. See BULLDOZER, MOTOR TRUCK.

screened well A cased well in which the water enters through one or more screens and not through holes in the casing. See OPEN-END WELL, PERFORATED-CASING WELL.

screens (intake works) Usually flat or round metal bars built into a frame, to keept out trash and debris, fitted at the entrance end of the structure for abstracting water from a river or other body of water. Facilities are provided for cleaning the screens. Fine screens may also be fitted in the INTAKE WORKS adjacent to the pump house. See BAND SCREEN.

scumboard A board which dips below the surface of a fluid so as to arrest the movement of scum.

sea A continuous volume of salt water which may form part of an OCEAN, e.g. Tasman Sea, North Sea, etc.

sea cliff A steep or vertical rock face formed by wave action. The height depends on the elevation of the land on which the sea advances, and the slope on the nature of the cliff rocks and rate of cutting. See WAVE-CUT TERRACE.

sea defence works Mainly structures and practices (1) to protect low-lying land against flooding by building walls or embankments, (2) to prevent or minimise erosion of the foreshore, (3) to limit the movement of sand and shingle by erecting groynes, and (4) to protect sand dunes by the planting of trees, grass, and shrubs. See SEA-WALLS.

seas See FORCED WAVES.

seasonal recovery (groundwater) The replenishment of water in zone of saturation during and following a wet season with rise in level of water-table.

sea-walls Protective walls built along the coast to arrest the force and erosive action of the waves; may consist of a low bank of sand, earth, or other material; agricultural land is often protected by banks on account of their relative low cost. They have a sloping seaward face covered with a pitching of stone blocks or other hard material. The wall may consist of concrete-filled bags, secured with steel spikes and surmounted by a parapet wall, the top of which is well above the level of the highest known tide. Walls of stone, concrete, or reinforced concrete, with a stepped or gently sloping seaward face are sometimes used, and also timber stakes and continuous timber barriers with shingle filling between them. The sea-wall may be constructed in conjunction with a system of groynes.

secondary main Generally, a medium-diameter pipeline forming a link between the TRUNK MAIN and SERVICE MAIN. It may be used to supply the larger consumers direct. See DUPLICATE MAINS.

second feet (U.S.A.) Rate of water flow in cubic feet per second or CUSECS. One cusec = 0·283 17 CUMECS.

second foot Rate of water flow equal to one CUSEC, or one cubic foot per second (U.S.A.).

second foot day Rate of water flow equal to one cubic foot per second for 24 hours. Equivalent to 86 400 ft^3 or 646 317 U.S. gal or 538 453 Imp. gal or runoff of 0·0372 in from 1 square mile.

section The representation, on a suitable scale, of geologic, hydrogeologic, or other features on a vertical or inclined plane. See CROSS SECTION (OF STREAM), STREAM PROFILE, STREAM SECTIONS.

sectional tank A water tank made from pressed-steel sections or units, commonly 1·2 m (4 ft) square with external flanges for bolting together. Usually 4·9 m (16 ft) is the maximum depth and in this case tie bars are used within the tank.

sector gate (or sector regulator) A type of ROLLER GATE in which the roller is not cylindrical but a sector of a circle. Usually the gate opens when moved upwards.

sediment (or silt) Rock particles transported by, suspended in, or deposited by a stream or other body of water; may be expressed as grammes per litre of water by weight or cubic centimetres per litre by volume. See LOAD.

sediment charge Ratio of volume or weight of silt to volume or weight of water passing a given cross section over a specified period.

sediment concentration The ratio of weight or volume of sediment in water to total weight or volume of the mixture; normally expressed in parts per million (ppm) or grammes per cubic metre (g/m^3) for low values of concentration and in percentage for high values.

sediment discharge The sediment transported in given period for given stage of a river, expressed in weight or volume units.

sediment discharge curve A graph showing relation between SEDIMENT DISCHARGE and stages of a river.

sediment hydrograph A graphic presentation of variation in SEDIMENT DISCHARGE or SEDIMENT CONCENTRATION with respect to time.

sediment runoff The sediment discharge in a specified period of a day, week, month, or year and termed daily runoff, weekly runoff, etc.

sediment spatial concentration The weight of sediment contained in a unit volume of stream or channel flow.

sediment transport competency See TRANSPORT COMPETENCY.

sediment transport concentration The amount of sediment transported compared with the amount of sediment and water carried by a stream past a given cross-section. See LOAD.

sediment water Water mixed with particles of mud, clay, silt or other material and obtained from, or existing in, a stream, canal, lake or other body of water. See SLIME, WASH LOAD.

seep Broadly applied to water which oozes out of the ground over a limited area; may indicate the outcropping of the water-table; a small outflow of water in a cutting or foundation.

seepage The percolation of water, generally outwards through the soil or rocks of an earthen channel or storage; the slow passage of water through a rock or soil mass. Seepage varies with the nature of the material, its condition, its position, and other factors. See EFFLUENT SEEPAGE, INFLUENT SEEPAGE.

seepage force See CAPILLARY PRESSURE.

seepage loss The volume of water lost by seepage through the banks or bed of a canal, earthen channel, or storage. Generally expressed in inches (or mm) loss in depth of water per 24 h. The loss varies with the nature and permeability of the ground and is difficult to measure accurately. The main methods are INFLOW-OUTFLOW, PONDAGE, PERMEAMETERS, and SEEPAGE METERS.

seepage meter An appliance for determining seepage loss in earthen channels; consists essentially of a watertight bell, a plastic bag, and a plastic connecting hose. Used mainly for detecting localised losses rather than losses over a considerable length of channel. See PONDAGE METHOD.

seepage spring A spring which occurs in sand or gravel in a valley or cutting located below the water-table, or along the upper surface of an impervious deposit. Seepage springs often occur along the line separating pervious and impervious rocks. They do not always flow continuously, but dry up in periods of drought or low rainfall. See FISSURE SPRIEG, VALLEY SPRING.

seep-off See SUBSOIL FLOW.

segment A specified portion of discharge section of a stream or channel, bounded by two consecutive verticals, the free surface above and the bed below.

segmental sluice gate See RADIAL GATE.

seiches Phenomena in which some types of earthquakes, with a long period between fairly large ground displacements, may initiate a relatively slow rocking motion, with appreciable magnitude, in the whole mass of water in a large reservoir or lake. The movement could cause considerable damage by the spillage of water over the top of the dam or over the banks of a lake. Also, it is considered that heavy winds or even sudden changes in atmospheric pressure may cause irregular variations or oscillations of the water level.

seismic method   A GEOPHYSICAL PROSPECTING method based on the speeds of transmission of shock waves through the ground to be tested. A shock wave is initiated by firing an explosive charge at a known point (shot) and recording the travel times for selected waves to arrive at receivers. The wave speed varies from about 600 m/s (2000 ft/s) for loose sediments up to about 6000 m/s (20 000 ft/s) for granite. Although limited in its application, it has been used to investigate the base of drift deposits and drift-filled channels. In this respect, relative depths are more reliable than absolute depths.

selected filling   Applied to the procedure of placing the finest and most adhesive of the bank-forming materials next to the core wall.

self-acting movable flood dam   A small timber dam AB (Figure S.1) pivoted to a back stay CD, the point C being placed at a distance of two thirds AB from the top. Hence when the water level is below A the CENTRE OF

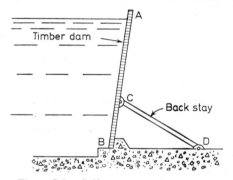

Figure S.1.   Self-acting movable flood dam

PRESSURE falls below C, and the dam is stable. If, however, the water flows over A, the centre of pressure rises above C, and the dam tips over. This occurs immediately during a flood. Each section has to be replaced when the flood has abated.

self-cleaning gradient   The slope of a pipe of a particular diameter at which the flow will carry away any solids normally contained in the water. Approximate gradients are 1 in 40 for a 100 mm (4 in) pipe, 1 in 60 for a 150 mm (6 in) pipe and 1 in 90 for a 230 mm (9 in) pipe. The gradient is often stipulated under bylaws affecting drains and sewers. It should not be too steep nor too flat.

self-docking dock   A type of FLOATING DOCK in which each section can be unbolted and lifted up on the others and repaired as required.

self-propelled hopper A self-propelled HOPPER BARGE.

semi-arid Describes a land or climate where the rainfall is greater than under arid conditions but still the amount limits the growth of many crops. Irrigation or dry-land farming methods are commonly required for crop production.

semi-diurnal tides Tides with two high waters and two low waters in each lunar day, that is, tides requiring roughly one half of a lunar day to complete their cycle. In general, the semi-diurnal tide is the most common. See DURATION OF TIDE.

semi-humid A climate or land which is somewhat intermediate between ARID and HUMID, with a tendency towards the humid condition.

semi-rigid joint A joint which permits limited movement in a pipe. The caulked lead spigot and socket joint is of this type and has been used in the U.K. for over a century. See FLEXIBLE JOINT.

separate system A drainage system in which sewage and rainwater are carried in separate sewers or drains. See COMBINED SYSTEM.

septic tank A tank in which sewage is subject to bacterial action and reduced to harmless substances. See CESSPIT.

service basin See SERVICE RESERVOIR.

service main Generally a small pipe from 75 mm (3 in) to 150 mm (6 in) diameter for conveying water to consumers in the local area or street in which they are laid. See SECONDARY MAIN, TRUNK MAIN.

service reservoir (or distribution reservoir or clear-water reservoir) A reservoir which forms part of a water distribution system and is connected to the main or IMPOUNDING RESERVOIR, often by GRAVITY MAIN; a storage of purified water to maintain supplies in case of failure of pumps or main supply line. A service reservoir supplies consumers in the vicinity with that part of their peak demand which exceeds the capacity of the pumps or filters.

set (1) The direction of flow of tide, current, etc. (2) A timber frame for supporting a tunnel.

settlement The gradual downward movement of an embankment or dam by gravitational compaction of soil and earth fill. Settlement of dam embankments will take place during and after construction. An allowance ranging from 20 to 40 mm per metre (0·25 to 0·5 in per foot) height of dam is usually made during construction so that the surface will eventually settle to the required level. Allowance for settlement is also made in the design of any structures built on the dam.

settling basin See DESILTING BASIN.

sewage The discharge from sanitary appliances attached to domestic, industrial, or other buildings. It flows down a drain connected to a public sewer or sometimes to a SEPTIC TANK.

sewage disposal The treatment or disposal of sewage by any method or process such as bacterial action, filtration, chemical precipitation, and also SEWAGE FARMS.

sewage farm A farm which takes sewage, usually as sludge, from settlement tanks; the sewage is used as manure and the effluent is drawn off to irrigate the land. A farm with a sandy soil is preferred for sewage disposal since sand is porous and acts as a filtering material. See BROAD IRRIGATION.

sewage gas (or **sludge gas**) A gas composed of two-thirds methane and one-third carbon dioxide and collected from tanks where sewage sludge undergoes DIGESTION.

sewer A large pipe, conduit, or channel usually covered over for carrying off sewage or water to the sewage disposal works. See SEPARATE SYSTEM.

sewerage An authority's planned system of sewers through which sewage is carried off to a disposal works. See COLLECTING SYSTEM.

sewer pill A wooden skeleton-framed ball which scrapes and cleans the sewer walls as it floats along; its diameter is slightly less than that of the sewer.

shadoof An ancient Egyptian device used to lift water from the Nile.

shaft-sinking pump A pump used for keeping a shaft dry during sinking operations. It is long, narrow, and designed to deal with about 4·5 m³/min (1000 gal/min) from the greatest estimated depth at which water will be encountered. It is often suspended from the surface, with facilities for raising and lowering when shotfiring at the shaft-bottom. Sinking pumps are usually of the electrically driven centrifugal type and allow for additional stages to be fitted as the depth increases. It may be suspended by a single-drum, worm-driven capstan engine with a very low speed. See SUBMERSIBLE PUMP.

shale A compacted clay or mud, containing from 70% to 80% of argillaceous minerals. Shale is an important deposit in groundwater engineering and when intact will not allow the passage of water, or only at a very slow rate. See CONFINING BED, PERCHED WATER-TABLE.

shallow manhole An inspection pit or MANHOLE in which the cross-sectional area is the same all the way up.

shallow percolation That part of precipitation that percolates downwards and laterally into streams. See DEEP PERCOLATION.

shallow well A well that taps water above the first impermeable rock layer; limited to a depth of less than 30·5 m (100 ft); usually lined with stone, brick, or concrete and the upper length made watertight. Periodic tests are desirable to detect possible contamination of water. Shallow wells are liable to seasonal fluctuations in yield and may dry up completely during a period of drought. See DEEP WELL, TUBE WELL.

sharpcrested weir A weir for measurement of discharge, consisting of a notch cut in a thin plate, having a sharp edge on the upstream side of the crest. See MEASURING WEIR.

sheathing A covering of sheet metal to protect underwater timber against MARINE BORERS.

sheepsfoot roller A tractor-towed ROLLER with projecting teeth about 230 mm (9 in) long used for compaction in COHESIVE SOILS. See SUPER-FICIAL COMPACTION.

sheet See NAPPE.

sheet erosion The wearing away of soil layers by RAIN-WASH, especially on sloping ground; a serious problem where the root-bound surface layers of a sandy soil have been removed. See ACCELERATED EROSION.

sheet flood A flood which has covered an alluvial fan or cone with a sheet of water. Also applied to a broad, shallow sheet of running water which appears and rises rapidly, generally after a cloudburst, and soon subsides and disappears—characteristic of some semi-arid regions.

sheet ice A sheet of ice formed near the banks of a stream, lake, or pond and extending slowly towards the centre; usually occur where the velocity is sluggish.

sheet piles Closely set interlocking units of wood, steel, concrete, or other material and forming a continuous lining as bank protection, or to keep loose earth or water out of an excavation; may also be used to stabilise permanent structures, foundations, and built cofferdams. See TIMBER SHEET PILING.

shelf ice A thick mass of fresh-water ice attached to, and extending from, the land. It may be afloat or aground and fed in part by glaciers; often found along a continental glacier or in large bays. See ICE CAPS.

Shell-perm process A process in which a permeable soil is injected with bitumen emulsion containing a coagulator to cause solidification in the soil voids. It reduces the flow of water into excavations. See GROUTING.

shell pump See SAND PUMP, BAILER.

shield A RAIN-GAUGE attachment which shields the gauge from wind and eddies, giving a more accurate rainfall figure.

shingle A beach or river deposit consisting of coarse gravel and pebbles of a size ranging from about 25 to 200 mm (1 to 8 in). See GRAVEL.

shingle bank A deposit or ridge of shingle formed along rivers and coasts where the water velocity is checked; sometimes found at river bends and along the lower reaches where the flow is too sluggish to carry the material further. See COMPETENCE (stream).

shingle-carrying rivers Rivers in which the material transported is mainly shingle and stone, derived perhaps from the erosion of steep slopes or landslides or faulting of beds. Transportation of shingle is active during floods and the material piles into heaps as the velocity diminishes. This causes the river to wander and erode its banks with widening of the channel.

ship caisson (or sliding caisson) A CAISSON for closing the entrance to a lock, dry dock, or wet dock; consists of a floating steel box fitting into a recess in the dock wall when the entrance is open or on to grooves in the walls of the entrance when it is closed. See GATE CHAMBER, STEP.

shoal A shallow; a sandbank exposed at low tide or water; a submerged bar across a river bed.

shooting flow See RAPID FLOW.

shore The zone or area between the highest high water line and the lowest low water line. See BACKSHORE, FORESHORE.

shore current A flow of water parallel to the shore line, formed by waves running inwards at an angle. The shore current carries material, in a zigzag fashion, across the shore. See LONG-SHORE DRIFT.

shore drift The coarse material covering the bottom along the shore in shallow water or where disturbance reaches the bottom. The material may include that delivered to the sea by streams or that produced by wave action. See LITTORAL DRIFT.

shore erosion The wearing away of the land belt adjacent to oceans and around lakes by the action of water, waves, and wind. See UNDERTOW, WAVE EROSION.

shore platform A wave-cut platform formed at the base of a headland and exposed at low tide; covered with sand and pebbles. See BEACH RIDGE.

shore protection Safeguarding a shore from the scouring effect of wave action and currents by the construction of BREAKWATERS, GROYNES, and REVETMENTS. See WAVE EROSION.

shot firing Increasing the supply of water in boreholes put down in sandstone, chalk, or limestone by detonating small charges of gelignite or dynamite to shatter and enlarge the fissures. Water increases of up to about 300% are possible if the explosives can be sited to give maximum results. See ACIDIFICATION.

shower A fall of rain or hail of short duration from convection clouds. See CONVECTIONAL RAIN.

shrinkage The decrease in volume of soil or fill material through the reduction of voids by mechanical compaction, superimposed loads, or natural settlement. See SUPERFICIAL COMPACTION.

side-channel spillway See LATERAL-FLOW SPILLWAY.

side contraction See CONTRACTED WEIR, END CONTRACTION.

side-entrance manhole A deep manhole in which the access shaft is located to one side of the INSPECTION CHAMBER which is reached by a passage from the bottom of the shaft.

side pond A storage pond at the side of, and connected to, a lock chamber. It reduces considerably the lockage loss.

side slopes See BATTERS.

Siebe diving helmet A diving helmet devised by Augustus Siebe in the 19th century and used by divers to the present day. A closed helmet is sealed to the watertight diving suit which is inflated by air to counterbalance the water pressure on the diver.

sight rail A horizontal board fixed at a specific height above a required level; used with BONING RODS for checking the levels set out for drains and trenches.

sill The horizontal overflow section of an irrigation check, or dam spillway, or measurement structure. Also applied to an underwater structure erected across a river to control or retain the level of the river bed. See LOCK SILL, NAPPE.

silt In hydrology, silt implies particles carried by, suspended in, or deposited by a river or other body of water. See SEDIMENT. In geology, it implies sediment composed mainly of fine sand and clay of dimensions between 0·02 and 0·002 mm. It becomes unstable in the presence of water and 'quick' when saturated. See SAND.

silt basin See DESILTING BASIN.

silt box A removable iron box fitted at the bottom of a GULLEY for collecting silt and grit. It is removed as required for discharge of silt. See GULLEY SUCKER.

silt-carrying rivers Those in which the material carried is mainly silt and sand. The rivers tend to have deep, well-defined beds, and meandering channels with wide floodplains. See SHINGLE-CARRYING RIVERS.

silt ejector See HYDRAULIC EJECTOR.

silt factor A coefficient which denotes the size of silt carried as LOAD by a stream or other body of water.

silt grade The average size of silt particles; sediment of silt size. See SILT.

silting The deposition of silt in a reservoir, river, sea bed, lake, or overflow area. Where it becomes excessive and obstructs flow of water, or of navigation, the material is removed by dredging or by groynes. The silting up of a reservoir is caused by inflowing streams depositing their sediment there. If the quantity is considerable, the water storage capacity of the reservoir may be seriously impaired in a small number of years. The remedies include washing the silt out through passages in the dam. Silt traps are sometimes constructed on the streams feeding the reservoir or a by-pass for flood waters around the reservoir. See SCOURING SLUICE.

silt jetties Long, narrow banks of silt on the sides of streams in digitate deltas.

silt sampler See BOTTLE SILT SAMPLER.

silt tank A small artificial basin, located on the upstream side of a farm or other reservoir, to retard the velocity of the incoming water and induce the deposition of silt and debris. The silt tank is connected to the main storage by concrete pipe or other conduit. See DESILTING (FARM STORAGE), DESILTING STRIP.

simple storage system A water conservation scheme, comprising a single reservoir and operated independently of any other water resources scheme. The system includes the reservoir, the catchment, and all necessary works, pumps, structures, and irrigation layouts, etc. See MULTI-STORAGE SYSTEM.

single-acting Applied to a RECIPROCATING PUMP in which water is delivered only every second stroke of the piston or plunger. See PUMP.

single-purpose reservoir A reservoir constructed to supply one specific demand, which may be water conservation, hydroelectric power, or flood protection.

single-stage pump A CENTRIFUGAL PUMP with a water lift of 30 m (100 ft). Normally, a six-stage pump would have a lift of 180 m (600 ft).

sink A water lodgment or trap at a pumping station. See SUMP.

sinker A plummet or weight attached to a CURRENT METER when taking velocities in streams. The sinker is usually of streamline shape.

sinker groyne A type developed in New Zealand and used mainly to retard the underscour of groynes. A stonemesh gabion, concrete block, or other heavy weight is placed at the outer end of the groyne and secured to a holdfast on the stream bank by wire cables to which obstructions, such as tree branches, are fixed. See TREE RETARDS.

sink-hole lakes Small lakes formed in cavities or SINKS which have become clogged with debris; usually found in areas where limestone underlies the surface soil. See CRUSTAL-MOVEMENT LAKES.

sinking pump See SHAFT-SINKING PUMP.

sinks Rough holes and cavities in the ground which join underground passages; usually occur in areas with limestone near the surface, portions of which have been dissolved by seepage of water. See KARST TOPOGRAPHY, SWALLOW HOLES.

sinuosity The meanders and loops in the course of a river flowing over relatively flat valley areas; the ratio between the length along the sinuous course of a river between two points and the straight-line distance between the two points.

sinuous flow See TURBULENT FLOW.

siphon A closed pipe and valves to conduct water from one level to a lower level on the opposite side of an intervening ridge of a height not exceeding about 7·6 m (25 ft). The difference in level between the inlet and outlet ends of the pipe column must be sufficient to give a motive head which overcomes the frictional resistance of the pipe and valves. The siphon was often used at an earlier period where power was not available. The term is sometimes used in irrigation to describe INVERTED SIPHONS.

siphon spillway A spillway which operates on the siphon principle; used in concrete dams where flood discharges are not very large and where space is limited. Sometimes called AUTOMATIC SIPHON SPILLWAY.

site The spot, locality, or area of land or water inspected, investigated, or selected for a dam, waterworks or wells. See INDEX MAP.

site investigations The collection of basic information on the geological structure and physical characteristics of the soils and rocks at the site of new water engineering projects or works; usually performed during the early planning stage. A project involving water supplies or storage would require information on the pervious and impervious rocks, on faults, and on the water-table. The putting down of trial pits or boreholes is often necessary. See DAM SITE INVESTIGATION.

site laboratory A laboratory located at a dam or other works in which tests can be performed to control fill placing, compaction, and so on. As a minimum it can perform moisture content, Proctor and *in situ* density, plastic and liquid limit, and shear strength tests. During the initial period, the laboratory is often used to establish standard procedures.

site plan A plan, prepared to a suitable scale, showing the location of a proposed dam or other project. It may show the land to be requisitioned, with ground, water-table, and perhaps rockhead contours. Fault lines, buried channels, and springs are usually shown. Each site plan is drawn to disclose information relevant to the project. See GEOLOGICAL MAP, INDEX MAP.

skimming To obtain relatively clean surface water by shallow overflow and holding back sand, silt, and debris carried as bottom load. Also applied to the levelling out of soil or surface irregularities.

skin friction See SURFACE DRAG.

slack water Water with no current or a very slight current; water about turn of tide, especially low tide.

slack-water navigation (or **still-water navigation**) Navigation in water channels with a very sluggish current. In a river the water may be made deeper with weirs and the river thus divided into reaches which are separated or connected by LOCKS. See SLUGGISH RIVER.

sleet A fall of rain mingled with hail or snow, or of snow melting while falling.

sliced blockwork BLOCKWORK for the construction of BREAKWATERS in which the stone or concrete blocks form sloping, nearly vertical courses. Each block is lowered by crane and slides into position much more easily than in COURSED BLOCKWORK. See TITAN CRANE.

slide See LANDSLIDE.

sliding caisson See SHIP CAISSON.

sliding gate A type of spillway CREST GATE with a vertical lift. It is difficult to raise against the water pressure and consequently its use is usually restricted to small spillways. See ROLLER GATE.

sliding-panel weir A FRAME WEIR consisting essentially of wooden panels which slide between a pair of grooved uprights.

slime Water containing fine particles which do not settle readily to the bottom; a fine oozy clay or mud. In the U.S.A. slime is defined as a suspension containing 50% or more of −200 mesh material or material below 0·0625 mm (1/400 in).

slip (1) See LANDSLIDE. (2) A stone or concrete structure which slopes down to the water's edge; a sloping plane on the bank of a river, used for ship-building; a contrivance for hauling vessels out of the water for repairs, etc.

slip circle The assumed circular arc mode of failure in a clay or dam embankment. See SLIP SURFACE OF FAILURE.

slip dock A dock with a sloping bottom from which water is excluded by a gate.

slip surface of failure (or **circular-arc method**) A form of SLIP in a dam, embankment, or other cutting when the ground is composed of homogeneous clay or similar material. The surface of failure closely follows the arc of a circle, which usually intersects the toe of the bank. Fixing the approximate position of the centre of rotation of the movement is important. See FELLENIUS'S CIRCULAR ARC METHOD.

slipway   A landing or shipbuilding SLIP.

slope   The gradient or inclination of a line or surface; expressed (1) in degrees from the horizontal plane, (2) as one vertical linear unit to a number of horizontal linear units, or (3) as the difference in elevation between two points divided by their horizontal distance apart. See BATTERS, GRADIENT.

slope gauge (1) An inclined staff or plank graduated according to slope, to give a more accurate depth reading than a vertical staff. (2) A gauge to determine water surface slope and fixed above and below DISCHARGE SECTION LINE of a stream. See STAFF GAUGE.

slope of stream   See STREAM GRADIENT.

sloping of banks   Bringing a river bank to a flatter slope and planting protective vegetation. This may suffice to check erosion if the flow is moderate and floods rare. The sloping bank is planted with strong rooting grasses and perhaps with shrubs or bushy trees along the lower part. The soil is often covered with brush until the grass has established itself. The lower part often needs additional protection against base erosion and caving of banks. See BANK PROTECTION.

slotted lining (boreholes)   A perforated or slotted lining sometimes used in boreholes in consolidated rocks where falls may occur. The perforations range from 13 to 75 mm ($\frac{1}{2}$ to 3 in) in diameter and formed at 75 to 300 mm (3 to 12 in) centres. The slots are from 150 to 300 mm (6 to 12 in) in length and from 6 to 20 mm ($\frac{1}{4}$ to $\frac{3}{4}$ in) in width. Similar slotted casings are used in loose ground with GRAVEL PACKING.

slough (1) A marsh, fen, swamp, or quagmire. (2) A secondary river with a sluggish flow (U.S.A.). (3) The breaking off or sliding of a sloping bank.

slow sand filter   A filter to remove particulate matter from a water supply and form a barrier against bacterial pollution. Usually a rectangular open basin, up to about 0·9 ha (2 acres) in area and built below finished ground level. The vertical or sloping filter walls may consist of brickwork, masonry, and clay puddle or concrete. The floor is paved with concrete or brick on clay puddle on which the under-drains are laid. About 0·6 m (2 ft) of successively finer gravel is laid over the drains, followed by 0·6 to 0·9 m (2 to 3 ft) of graded sand. The use of slow sand filters is declining, largely on account of high cost. See FILTER.

sludge gas   See SEWAGE GAS.

sludger   See BAILER.

sludge sample   The broken material produced during drilling and sometimes used as a sample of the rock passed through. See CORE.

211

sluggish river A river with little flow or current, in which peaks of flood form more slowly because of low slope or because of retarded or reduced current by heavy withdrawal or storage in upper reaches of a river system. A very sluggish river may have only fine silt as its bed load. See OLD RIVER.

sluice A water-gate or floodgate; a channel for passage of water fitted with a vertically sliding gate for controlling the flow; an opening in a structure for passing water at high velocity and for discharge of silt and debris. See SCOURING SLUICE.

sluice box (irrigation) A structure made of steel, wood, or other material to enable the irrigator to control the flow of water in each bay. Also called bay outlet or ditch outlet.

sluice valve A valve in which a bronze spindle operates a solid tapered slide with gunmetal faces. It is not a regulating valve and when so used it may be seriously damaged. See VALVE.

sluicing A term sometimes used for HYDRAULICKING or HYDRAULIC EXCAVATION. See HYDRAULIC MINING.

slushing See HYDRAULIC FILL.

small-area floods The flooding of a relatively small area by high-intensity storms with a duration of up to 24 h. See LARGE-AREA FLOODS.

smoke test A test carried out with smoke in a new pipeline to locate any leakages, or to assess the efficiency of drains.

smooth-wheel roller A self-propelled ROLLER for earth compaction. See COMPACTION PLANT, SUPERFICIAL COMPACTION.

snag A stump or trunk of a tree which forms an obstruction in a river.

snagging The removal of obstructions from the channel of a stream, particularly fallen trees. The clearing of a river channel lessens the risk of overflows and minor floods, and is often one of the first steps in improving a river. For removing snags, tractors are often used, particularly those with winches attached (Australia). See PREVENTATIVE SNAGGING.

snifter valve A valve designed to admit a small quantity of air into a pipeline to keep an air vessel charged with air.

snore A pump suction end which is submerged in the SUMP and through which water passes to the pump at the upper end of the pipe. It may be a cast-iron cylinder, with an oval or flat-bottomed STRAINER. The water entering the pump is thus relatively free from dirt and debris. See FLOATING STRAINER.

snout The projecting end of a glacier, usually in an area below the SNOW LINE.

snow Atmospheric vapour frozen into ice crystals of needle-like or feathery structure and falling to earth in white flakes or spread on it as a white mantle; the flakes may consist of a number of single crystals matted together. See below.

snow course A line permanently marked on the SNOW COVER, along which snow samples are taken at suitable distances and periods to determine depth and SNOW DENSITY. The measurements enable forecasts being made of the subsequent runoff. See SNOW SURVEY.

snow cover (or **snow mantle**) A layer of snow and ice which covers an area for a considerable period.

snow density The water content of snow, expressed as a percentage by volume. In sampling a snow cover, it is the ratio of the scale reading in millimetres (or inches) of water to the depth in millimetres (or inches) of the snow core. See SNOW SAMPLER.

snowdrift A bank of snow formed in depressions and sheltered places by the action of wind, or by snowfalls in strong wind.

snowfence A barrier or fence to impound snow in areas where melting in place contributes to soil moisture; a fence of slat and wire or other construction to intercept and hold drifting snow and thus safeguard engineering works, railways, and roads.

snowfield A large tract covered with snow, usually in mountainous or polar regions. See SNOW LINE.

snow-gauge An instrument for measuring the amount of snow that falls during one storm or over a specified period at one place. See SNOW SAMPLER.

snowhedge A fence or barrier of bushes, shrubs, or low trees planted to intercept drifting snow.

snow line The lower limit or line in altitude of perpetual snow on mountains, varying in height according to climate. In equatorial regions, the line may be as high as 5500 m (18 000 ft) above sea level, while in polar regions it is below sea level.

snow mantle See SNOW COVER.

snow melt The water formed when snow melts; the water may seep downwards, become part of surface runoff, or evaporate. See EFFECTIVE SNOW MELT.

snow sample A core extracted by means of a snow sampler from a point on the SNOW COURSE to determine snow thickness and density.

snow sampler A device consisting of light jointed tubes for taking snow samples. A spring scale gives a direct reading of the depth of water contained in a given sample of snow. See SNOW-GAUGE.

snow stake A vertical stick, prop, or post, generally permanent, which indicates the depth of snow in areas with a thick snow cover.

snow survey The sampling of the snow cover at selected points to determine snow depth, density, and water content. The water stored as snow on a catchment area and the subsequent runoff may be estimated from the survey data.

snubber A fitting used for the protection of pressure gauges against hydraulic shock and sudden changes of pressure. See HYDRAULIC PRESSURE SNUBBER.

soakaway See DRY WELL.

213

sod strips   Strips or bands of sod or turf left in a natural water channel to prevent erosion; narrow ridges or bands of grass or other close-rooted crop spread across a channel to retard the flow of water. See MATTRESS.

soffit   The uppermost part of the internal surface of a culvert, drain, sewer, or channel (Figure S.2). See INVERT.

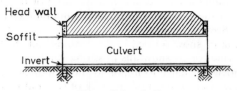

Figure S.2.   Soffit

soft water   Relatively pure water which lathers readily when mixed with soap. See HARDNESS OF WATER.

soil   The upper layer of earth on which rain falls and plants grow. It is composed of mineral particles, some organic matter, and water and may range from fine clay or mud to gravel or boulders. To the engineer, soil is the superficial material that is not hard or solid rock. Soils may be classified into certain well-defined types, depending mainly on the size and nature of the particles, as below.

Grade	Dominant grain size
Gravel, scree, shingle	2 mm and over
Very coarse sand	2 mm to 1 mm
Coarse sand	1 mm to 0·5 mm
Medium sand	0·5 mm to 0·25 mm
Fine sand	0·25 mm to 0·1 mm
Silt	0·1 mm to 0·01 mm
Clay or mud	Less than 0·01 mm

soil conservation   The management of land and soil, including drainage or irrigation where necessary, to obtain the maximum profitable production of crops and stock on a sustained-yield basis. See DRAINAGE, IRRIGATION, SOIL EROSION, SUBSURFACE DRAINAGE, WATER PROSPECT MAP.

soil conservation survey   A report and a map or aerial photograph showing the physical land features and associated data in sufficient detail for planning soil conservation and land usage measures.

soil conserving crops   Grasses and other crops that prevent or retard erosion and maintain or replenish soil organic matter. These crops, when cut for hay, yield considerable aftermath or after-grass.

soil discharge   The loss of water from the ZONE OF SATURATION through evaporation directly from the soil or rocks. It occurs only in areas where the water-table is very close to the land surface. See EVAPORATION DISCHARGE.

soil erosion The wearing away and removal of the soil layers mainly by the action of wind, rain, and running water. The process may remove the entire mantle of soil, making the land valueless for cultivation; it sometimes occurs in areas of over-cultivation and ill-planned deforestation. Soil erosion often creates problems for engineers concerned with floods and silting in reservoir basins, rivers, and canals. The remedies include control of over-grazing, afforestation on the steeper slopes, contour trenching or ridging, construction of barriers, and weirs, or detention dams, etc. See GULLY DRAINAGE, SHEET EROSION.

soil evaporation The evaporation of water from the surface and near surface soil layers.

soil grades See SOIL.

soil mechanics The systematic study of the nature, composition, classification, consolidation, and related properties of soils. This information determines their suitability for dams, canals, reservoirs, and other water engineering works. Dr Karl von Terzaghi, M.I.C.E. (1883—1963), is generally regarded as the founder of modern soil mechanics. See DAM FOUNDATION, DAM SITE INVESTIGATIONS, SITE INVESTIGATIONS.

soil moisture (or **soil water**) Water occurring in the voids of the soil mantle and is gradually discharged by evaporation or by transpiration of plants. See BELT OF SOIL WATER.

soil moisture deficit The quantity of water, expressed in millimetres or inches required to restore a soil to FIELD CAPACITY. In general, plants growing in heavy (clayey) soils can withstand higher deficits than plants growing in light (sandy or coarse-grained) soils. Therefore, plants in sandy soils need higher rainfall or more frequent irrigation than plants in clayey soils. See PERMANENT WILTING POINT.

soil moisture suction The pressure (below atmospheric pressure) which causes water to adhere or be retained in soil or rock above the zone of saturation. See CAPILLARY WATER, ZONE OF AERATION.

soil probe A pointed steel rod sometimes used to measure the depth of water penetration in soil irrigation. See PROBING.

soil-saving dam A ridge or barrier of earth constructed across a stream or gully to impound runoff and silt; used to minimise erosion and promote the deposition of eroded soil material. The term is often restricted to dams which are larger and of a more permanent nature than CHECKDAMS.

soil-saving dike An earth ridge formed along the lower end of an irrigation field, often with an adjustable outlet; used mainly to conserve soil or sediment within the area (U.S.A.).

soil structure The arrangement or grouping of the primary soil particles into soil aggregates, which may be coarse, medium, or fine. Soil structure affects the rates of absorption and movement of water.

soil water See SOIL MOISTURE.

215

solid map   A map showing the solid rocks of an area and assuming that all the superficial incoherent deposits, other than alluvium, were absent. See GEOLOGICAL MAP.

solids-handling pump   Commonly a centrifugal pump for dealing with water or liquids containing sand, ore tailings, or fine coal; it is designed to resist abrasion and rubber linings are often used, as they last longer than iron or steel. See PULSOMETER PUMP.

solid wharf   See WHARF.

soling   Large stones in PITCHED WORK.

solution cavities   Irregular underground passages and pits formed in certain soluble rocks, such as limestone, by the seepage and flow of water. In some cases, percolating solutions have filled the cavities with minerals. See SINKS.

solution channel spring   A type of spring often found in limestone districts; water flows along channels enlarged by solution and emerge as a spring in a valley. Often seen in the limestone districts of the Pennines and the Mendips (U.K.).

solution valley   A valley formed by the chemical wearing away of soluble rocks; often broadly 'U' shaped in cross-section.

sorting   Sometimes applied to the progressive reduction in size of particles forming the LOAD of a stream as it flows towards the sea or other body of water.

sounder, echo   See ECHO SOUNDER.

sounding   Determining the depth of water in a river, ocean, or reservoir at a particular spot by sending out a sonic or ultrasonic sound impulse and measuring the interval between the transmission and the reception of an echo. For accuracy, many factors must be known, such as the type of water, etc. The term is sometimes used in measuring the depth to bedrock, or thickness of soil, by forcing a steel rod into the soft ground.

sounding lead   Sometimes applied to a SOUNDING WEIGHT.

sounding line (or lead line)   A strong cord up to about 180 m (100 fathoms) in length, knotted at fathom intervals, and used for taking SOUNDINGS in hydrography. A weight is attached to take the line downwards to the bed of a river, lake, reservoir, or sea.

sounding line corrections   The adjustments made to the sounding line measurements to allow for departures of the line from the vertical. See AIRLINE CORRECTION.

sounding rod (or sounding stick)   A graduated rod or stick weighed with lead at one end and used for determining the depth of relatively shallow water, such as streams and canals.

sounding weight (or lead)   See SOUNDING LINE.

sounding wire   See LOG LINE.

sowing with grass seed   A dam contract term which includes sowing, watering, fertilising, and maintenance during the construction period. Measurement is in square metres or square yards.

spall drain   See RUBBLE DRAIN.

spatial concentration of sediment Sometimes applied to the sediment contained in a unit volume of flow. See SEDIMENT TRANSPORT CONCENTRATION.

specification A document giving a minute technical description, or enumeration of particulars, materials, construction, and workmanship of any hydraulic work to be carried out under the contract.

specific capacity (of well) The rate at which water can be pumped from a well per unit of DRAWDOWN; only useful where the drawdown varies approximately as the yield from a well. It may be expressed in litres per minute per metre (drawdown) or gallons per minute per foot (drawdown).

specific discharge A quantity often used for measuring and comparing floods and equals discharge per unit area.

specific energy A value of flow energy in a conduit or open channel, with the bottom of channel as base for measurement; expressed as energy head above base of channel.

specific gauge reading Water level at a specific station for a given discharge.

specific gravity (s.g.) The ratio of the weight of a given volume of soil, rock, or other substance to the weight of an equal volume of water at a given temperature.

specific retention The ratio of the weight or volume of water which a soil will retain against the force of gravity to its own weight or volume; determined after soil has once been saturated. See EFFECTIVE POROSITY, SOIL MOISTURE SUCTION.

specific yield The amount of water which a unit volume of soil or rock will yield after being saturated and allowed to drain under specified conditions; expressed as percentage of volume. See SPECIFIC POROSITY.

speedy moisture tester An instrument for the rapid determination of moisture in a soil sample. It uses calcium carbide and a pressure gauge is calibrated to give direct values of moisture percentage.

spill Overflow of water in tidal estuaries during rise of tide or of river banks during high floods.

spillover That part of OROGRAPHIC PRECIPITATION which reaches the ground on the lee side of the mountain or barrier, having been carried forward by horizontal wind.

spillway (dams) A structure which passes floodwater through, over, or around a dam. Provision is usually made to prevent scour and the removal of rock at the downstream face of the dam by spillway discharges. This may consist of a concrete apron at the toe or other constructions. In the design of spillways, rainfall records, particularly of past heavy storms, are important. See APRON, BY-WASH, FLASHBOARD, OVERSHOT SPILLWAY, ROLLWAY.

spillway capacity The capacity of a dam spillway to pass flood water. For example, spillways for large farm dams should be capable of carrying the one in a hundred-year flood. See FLOOD (RETURN PERIOD).

spillway crest level The upper level of crest structure which regulates water flow in a spillway (Figure R.1). See FLASHBOARD.

spillway gate See CREST GATE.

spillway overflow See OVERFLOW SPILLWAY.

spin test Checking whether a CUP TYPE CURRENT METER is in good condition by blowing into the cups and spinning the rotor. See MINIMUM SPEED OF RESPONSE.

spit A ridge of land or deposit of sand or shingle projecting from headlands and terminating in the open sea; a narrow and submerged ridge across the mouth of a bay. See BAY-MOUTH BARS, HOOKED SPITS.

splash erosion The loosening and erosion of soil particles and soil layers by the action of raindrops. See RILL EROSION.

sporadic permafrost Scattered islands of PERMAFROST in ground which is predominantly unfrozen.

spout-delivery pump A pump which cannot deliver water above its own height and is similar to the DIAPHRAGM PUMP.

spray irrigation See SPRINKLER IRRIGATION.

spraylines A method of irrigation often used on rectangular plots from 9 to 15 m (30 to 50 ft) in width according to water pressure. Spraylines consist of galvanised steel, aluminium, or plastic tubing of 25 to 50 mm (1 to 2 in) internal diam., with quick-couplings to ease moving. The tubing is fitted with nozzles spaced at about 0·6 m (24 in) and designed to give the desired output and spray pattern. The spacing of the spraylines is related to the throw of the water jets. See TRICKLE IRRIGATION.

spreader (1) Any appliance which distributes water uniformly in or from a channel. (2) A machine fitted with wide blades for spreading soil, subsoil, or rock excavated from pond, drainage ditch, or other cut.

spreader strip A plot of land, of variable width, planted to a erosion resistant crop, to regulate and fan out the runoff from cultivated land on the rise side. The strip is planted across the general slope. It helps to distribute runoff and control erosion. See BORDER DIKES.

spring A natural outflow of groundwater at the surface. Springs often serve as useful sources of water. Exceptionally, seven of the Havant and Bedhampton Springs of the Portsmouth Water Company (U.K.) never yield less than 4500 m³ per day (1 m.g.d.) each. See ARTESIAN SPRING, BOURNES, CONTACT SPRING, FAULT SPRING, FISSURE SPRING, THERMAL SPRING, TUBULAR SPRING, VALLEY SPRING.

spring range See MEAN TIDAL RANGE.

spring tides The highest ordinary tides that occur at, or soon after, the new moon and the full moon. See EQUINOCTIAL TIDES, NEAP TIDES.

sprinkler irrigation The spreading of water over soil by means of sprays from pipes or pipe projections above the ground level. Rotating sprinkler heads are often used with one or more nozzles which discharge the jets. The sprinklers may each water a circle of from 7·6 to 13·6 m (25 to 45 ft) radius, and be spaced from 6 to 10·6 m (20 to 35 ft) intervals along the lateral line according to the planting pattern and other local factors. See SPRAYLINES.

spud (or **anchoring spud**) A steel post under a DREDGER which provides anchorage during dredging operations. The post is lowered by ropes or a toothed rack until it is secured in the river or sea bed.

spur (or **wing dam** (U.S.A.)) (1) An artificial obstruction formed from the bank of a channel and projecting into it for training the flow of water and for bank protection; a similar structure for protecting a shore from erosion. A spur may be constructed of stone, earth, timber, or brushwood; the head is reinforced to resist scour. (2) A minor topographic feature, such as a small hill extending from a large dominant range of mountains.

spur dyke See SPUR.

spur terrace A short ridge of earth for impounding or diverting runoff.

squall A sudden and violent gust or successive gusts of wind, especially with snow, sleet, or rain.

square check irrigation A modification of CHECK IRRIGATION in which rectangular or square basins are used to confine irrigation water.

square-mile foot A volume of water one foot deep covering one square mile and is equivalent to 789 440 cubic metres. See ACRE FOOT.

stability (1) In a channel carrying water stability is established when ACCRETION balances SCOUR. (2) The resistance of a river bank or valley slope to sliding or deformation; depends largely on the shearing strength, the water content of the material, and the degree of slope. See BATTERS, CAVING.

stability number A number pertaining to the stability of a soil embankment. Thus:

$$\text{stability number} = HW/C$$

where H is the critical height of sloped bank, W is the effective unit weight of soil, and C is the cohesion of soil.

stabilised grade That grade or slope of a stream or channel along which neither erosion nor silting occurs. See PROFILE OF EQUILIBRIUM, REGIME.

stack A high detached rock rising out of the sea and near the coast (U.K.). See CHIMNEY.

staff gauge A vertical staff, plank, rod, metal plate, wall of a pier, etc., with a graduated scale by which depth of water in a river, dock or canal may be read. See INCLINED GAUGE, SLOPE GAUGE, SOUNDING ROD.

stage (or **gauge height**) Elevation of water surface, usually in a stream, relative to a fixed datum or plane of reference. See BANKFUL STAGE, BED BUILDING STAGE, FLOOD STAGE, RIVER STAGE, WARNING STAGE.

stage discharge relation The relation between water level and discharge of a conduit at a specific station for a given condition of flow. See DISCHARGE CURVE.

stage profile See STREAM PROFILE, SURFACE PROFILE.

stage recorder See WATER STAGE RECORDER.

stage relation See GAUGE CORRELATION.

stake and stone sausage construction A river protection method in which the bank is sloped and on the face vertical willow stakes are placed. Over these, long willow stakes are fixed horizontally near the top of the bank. Stone sausages (see SAUSAGE CONSTRUCTION) are then laid at about 3 m (10 ft) centres down the bank slope and anchored by wires to posts fixed well back from the bank. See FASCINES.

staking (or **pegging out**) Fixing the location or line of drains, channels, terraces, ponds, and similar structures and the placing of metal or wood stakes as markers; staking a drainage ditch includes marking the line and also the depth or slope of the ditch.

standard current meter A calibrated current meter used for checking and comparing the rating of other current meters. See REFERENCE CURRENT METER.

standard flood Flood data used in certain equations in which it is assumed that the duration of the storm is equal to the PERIOD OF CONCENTRATION; the storm is stationary and its area just covers the catchment; the rainfall is of constant intensity over the catchment and throughout the storm; and the catchment has a constant slope and runoff coefficient.

standardisation An agreement between authorities, manufacturers, and engineers on certain qualities, tests, dimensions, and tolerances of a certain practice, product, part, or machine and to adopt same in engineering practice. It offers many advantages such as interchangeability of machine parts and a reduction in stocks of spares. In the U.K. there are agreements between producer and user under the authority of the BRITISH STANDARDS INSTITUTION.

Standard Method of Measurement A procedure recommended by the Royal Institution of Chartered Surveyors for measuring structures.

standard mix CONCRETE mixed in the volumetric proportions of 1 cement, 2 sand, and 4 of coarse material.

standard project flood A flood representing a synthesis of flood occurrences in a given area. Its frequency is unspecified, but it forms a sort of datum or standard against which other floods are compared. See DESIGN FLOOD.

standard system of levels The system of levels used at gauging stations and other sites and related to a recognised datum, such as the ORDNANCE DATUM.

standing water level (or **rest water level**) The water-table when left undisturbed for a period; in general, the level at which water stands in wells when no pumping has taken place. See FREE WATER LEVEL.

standing wave flume A flume with bottom contractions or side contractions or both and within which the velocity changes from SUB-CRITICAL FLOW to HYPER-CRITICAL FLOW. The discharge is measured by the velocity of flow at critical depth within the throat and the cross-sectional area.

standpipe (1) A pipe projecting vertically upwards from a water main under pressure. It may be left open at the top and the water level will indicate the pressure in the water main; sometimes closed at the top with a draw-off valve. See SURGE PIPE, WATER HAMMER. (2) A circular reservoir of reinforced concrete or steel constructed above ground level.

stank (1) To seal off or make watertight. (2) A small timber COFFERDAM sealed with clay. Formerly applied to a pool, ditch or tank.

starling A series of timber piles driven into a river bed on the upstream side of a bridge pier, and also downstream in tidal rivers, to protect the structure from floating ice blocks and debris.

static delivery head The vertical height, usually in metres or feet, between the pump centre and the outlet of the delivery pipe.

static head The head produced by static pressure of water; equal to the pressure divided by unit weight of water. The difference in water levels between two locations, such as the impounding reservoir and the service reservoir; in a pump, the sum of the static suction head and static delivery head.

static level See HYDROSTATIC LEVEL.

static suction head The vertical height, usually in metres or feet, from the free water level in the sump to the centre of the pump. See SUCTION HEAD.

stationary dredge See DREDGER.

staunching piece (or **staunching bead**) The vertical gap left in a concrete dam between successive bays. The concreting of the gap is carried out after the shrinkage of the adjoining bays.

staunching rod A strong rubber rod used at a CREST GATE to form a watertight joint; the rubber is compressed between the gate and the dam structure.

steady flow See STREAMLINE FLOW.

steening (or **steining**) The lining of a well with bricks or stones which may be laid with mortar but are usually dry. See DRY WELL, WELL SINKER.

step Bringing a LOCK GATE to the vertical position; a gate is usually towed to the site in a horizontal position and stepping is slow and awkward.

step irons (or **foot irons**) Staples of galvanised malleable cast-iron built into the brick or concrete wall of a MANHOLE to provide access to a deep sewer.

sterilisation To render water free from undesirable bacteria and therefore safe for human consumption. The boiling of all water is the most positive and readily available method of sterilisation. See CHLORINATION.

Stevenson's formula A British formula for calculating the height of waves developed by wind action over a water distance of F nautical miles (the FETCH). The height in feet is equal to $1 \cdot 5 \sqrt{F}$. The rule applies only to unobstructed deep water, and for distances of 400 miles (640 km) or more the values given are excessive. See FETCH, FORCED WAVES, HAWKSLEY FORMULA, WAVE HEIGHT, WAVE LENGTH, WAVE PRESSURE.

stiff-fissured clay A clay which is firm when dry at depth but is intersected by numerous cracks along which water can seep readily.

stilling basin See STILLING POOL.

stilling pool (or **stilling basin** or **water cushion**)   A water basin formed at the foot of a dam spillway or waterfall or rapids to reduce the velocity and scouring action of the descending water. See APRON.

stilling well   A well in which the water enters through small inlets in a chamber or pipe; these dampen the waves or surges while allowing the water level within the well to rise or fall steadily with the movements of the main body of water. See STRAINER WELL.

still-water navigation   See SLACK-WATER NAVIGATION.

stirrup wire   The wire connecting the lower and upper sheets of wire mesh in MATTRESSES and other river protection works.

stockade groynes   A modified form of the CRIB GROYNE with shorter timbers. A box structure is formed, with open top and bottom, by saplings strung vertically on steel cables and filled with stone or shingle. The construction is stiffened by cross ties and braces. See CRIBWORK.

stock pond   See FARM POND, TANK, EARTH.

stockwater development   The development in rural areas of new and improved schemes of water supply for stock. It includes investigations covering water sources, both surface and underground, and the design of associated delivery conduits and storages. See SURFACE WATER SUPPLIES.

stone, angular   See ANGULAR STONE.

stone apron   See UNDERWATER APRON.

stone drain   See RUBBLE DRAIN.

stone-filled work   All shore, river, canal, and dam works in which stone (angular or rounded) forms an essential construction material. See BEACHING, GROYNES, RIPRAP, SAUSAGE CONSTRUCTION.

stone, loose   See LOOSE STONE.

stonemesh construction   A form of river bank protection consisting of stones 50 to 150 mm (2 to 6 in) in size enclosed in a wrapping of wire mesh. As the wire will gradually rust and fail, the construction is viewed as semi-permanent. It should, however, hold for a few years until the establishment of grasses or willows to bind the stone together. See STONE RIPRAP.

stonemesh groynes   These groynes are used mainly to divert the river flow away from an eroding bank; used extensively in New Zealand in the smaller rivers, particularly where rounded stone is plentiful and where timber suitable for cribwork is not available. They consist essentially of rounded stones held in a wrapping of wire mesh, and are often built up with rectangular gabions laid in courses; the construction tapers as the work proceeds to the top. See GABIONS.

stonemesh mattress (or **stonemesh apron**)   A type of STONEMESH CONSTRUCTION used where protection of beds or banks of rivers from erosion is of special importance. It may be used around bridge piers, at the roots and heads of groynes, and below weirs, etc. It consists essentially of a layer of stones, some 300 mm (12 in) thick, with a layer of mesh right around to form a mattress or apron. Should scour commence to undermine the construction, the weight of the stones causes the apron to sag and adapt itself to the new surface, while the stones are held in position and cannot be rolled away. See STONE RIPRAP, UNDERWATER APRON.

stone paving A hard floor or paving consisting of roughly flat stones, hand-placed on the downstream side of a river structure to minimise scour of the bed. See APRON.

stone pitching A facing of stones, up to 0·45 m (18 in) or more in thickness, above the BEACHING on the WATER FACE of an earth dam as a protection against erosion and wave action. Sometimes precast or *in situ* concrete is used.

stone riprap (or **beaching**) The use of loose, angular stones to check river bank erosion; used particularly in areas where good hard stone can be obtained cheaply. The stones should be durable and each of a weight sufficient to resist transportation by the strongest current likely in that river. If the water is deep and the eroding bank nearly vertical, it may be cheaper to use extra stone instead of sloping the bank. Greater protection is provided by placing a layer of gravel or small stones against the bank and then building up the large stones over it. See BEACHING, SLOPING OF BANKS.

stoning Preventing scour along a river bed by means of a layer of broken stone. See REVETMENT, RIPRAP.

stop See BAY OUTLET.

stopbanks New Zealand term for LEVEES.

stop-log See FLASHBOARD.

stop-plank See FLASHBOARD.

stop valve A valve to stop or start the flow of water, usually in a pipe. See DISCHARGE VALVE.

storage The impounding of water in underground or surface reservoirs, tanks, canals, rivers, or basins and from which supplies may be drawn as required. See ARTESIAN BASIN, BANK STORAGE, CHANNEL STORAGE, CONSERVATION STORAGE, DEPRESSION STORAGE, GROUNDWATER STORAGE, PONDAGE, RESERVOIR, VALLEY STORAGE.

storage balance See PONDAGE.

storage canal See CANAL, MAIN IRRIGATION.

storage capacity The maximum volume of water which a reservoir can hold without overflowing; may be expressed in cubic metres, cubic yards, millions of gallons, or acre feet. The required capacity of a FARM RESERVOIR depends on the approximate water requirements for domestic, stock, and irrigation and includes estimated evaporation and seepage losses. Where the borrow pit is inside the natural reservoir basin, the total capacity equals the natural storage plus the volume of the borrow pit. See PONDAGE, STORAGE/EXCAVATION RATIO.

storage coefficient (aquifer) The ratio of (1) the volume of water taken into, or released from, storage in a prism of aquifer of unit surface area and the total thickness of the aquifer to (2) the volume of the prism of aquifer per unit change in the component of pressure head normal to that surface.

storage curve   See CAPACITY CURVE.

storage curve, groundwater   See GROUNDWATER STORAGE CURVE.

storage cycle   Total time or period at end of which the level or volume of water in reservoir is same as at commencement; may be a few hours, 24 hours, one year, or a longer period.

storage, economic   See ECONOMIC STORAGE.

storage/excavation ratio (or **storage ratio** or **S/E ratio** or **water-to-earth ratio**)   The ratio of volume of water stored to the volume of earth excavated for the bank or dam. Thus a storage ratio of 10:1 means that the volume of water is ten times the volume of earth excavated. See DAM SITE INVESTIGATIONS, STORAGE CAPACITY.

storage of aquifer   The quantity of water released from storage in aquifer with a given lowering of head.

storage period   The period without water inflow to the reservoir. Over such a period the consumer requirements must be met wholly from the water held in storage.

storage, pumped   See PUMPED STORAGE (HYDROELECTRIC), PUMPED STORAGE (RESERVOIR).

storage ratio   See STORAGE/EXCAVATION RATIO.

storage regulator   See PONDAGE.

storage reservoir   A reservoir for storing water as part of a flood protection scheme or for water supplies. See SERVICE RESERVOIR.

storage routing   A method of FLOOD ROUTING by determining the actual volume of inflow to, outflow from, and storage in the reservoir.

storage, surface   See INITIAL DETENTION.

storm   A violent disturbance of the atmosphere with heavy rain, or snow or hail, and often thunder and lightning; a violent gale with rain or snow. Storms vary greatly in duration, in intensity, and in movement; some are part of seasonal conditions and lasting for months, while others are isolated phenomena; some move rapidly while others remain almost stationary for days and cause heavy local precipitation and floods. See HAILSTORM, MONSOON, PRECIPITATION, EXCESSIVE, THUNDERSTORM.

storm distribution pattern   A plan, chart, or layout showing variation in depth of rainfall over an area for a specific storm.

storm flow   Excess water flowing after heavy rain, at or near the surface, into streams or rivers.

storm flow, subsurface   See SUBSURFACE RUNOFF.

storm lane (or **storm belt**)   The belt or tract in which storms are frequent.

storm pavement   A gently sloping paved bank to a BREAKWATER.

storm rainfall   Rainfall associated with a given storm; it may affect many thousands of square kilometres (or square miles) and vary in duration from a few minutes to several days.

storm runoff   The runoff after a storm or snow melt, which may reach a measurable amount soon after the occurrence of rain or thaw.

storm sewer A SEWER which normally has no flow but carries water discharged after heavy rain.

storm wall See PARAPET WALL.

stormwater (1) The water which drains off a catchment area during and after a heavy fall of rain or snow. (2) In a combined system of drainage, the term is often applied to a stated excess of drainage water above the dry-weather flow at which the sewage is permitted to overflow into a stream after passing through STORMWATER TANKS. In general, before the sewage overflows into STORM SEWERS the total flow must be three times the average dry-weather flow. See COMBINED SYSTEM, PARTIALLY SEPARATE SYSTEM.

stormwater tanks Cisterns or vessels in which sediment and solid material settles from stormwater before it passes into a stream. See SILT BOX.

storm wave The overflow of low coasts, not normally flooded, by TIDAL CURRENTS and sometimes strong wind. See TIDAL WAVE.

strainer A perforated pipe, cylinder of wire-gauge, or similar tube. It allows the passage of water but not large particles. Used in RADIAL WELLS, STRAINER WELLS, and PUMPS. See FLOATING STRAINER.

strainer well A well in which a strainer is positioned in the borehole where it intersects the aquifer to exclude solid particles above a certain size. See STILLING WELL, WELL SCREEN.

stranded caisson See BOX CAISSON.

stratigraphic bores Boreholes put down primarily to establish geological sequence in a given area. In hydrogeology, the position, structure, and continuity of the pervious and impervious deposits and faults would be major objectives. See STRUCTURE BORES.

stratum spring See CONTACT SPRING.

stream (or brook or rill or rivulet) A general term for a flow of water in a natural surface channel; a flow of water in an open or closed conduit; a jet of water issuing from an orifice or notch. See BROOK, POTAMOLOGY, SUBSEQUENT STREAM, SUBTERRANEAN STREAM, WATERCOURSE.

stream at grade The slope along the course of a river, below which further erosion is no longer possible. When a river has cut its channel down to this critical slope it is said to be 'at grade'. When at or near grade a river is said to be mature, and there are no irregularities such as pools and waterfalls. See GRADED STREAM.

stream bank See BANK, RIGHT BANK, RIPARIAN.

stream bank erosion The removal of soil and rocks along the banks of a stream by scour and caving of overlying deposits; some bank erosion also occurs by runoff from adjoining land, especially with steep slopes.

stream bank erosion control The use of mechanical or vegetative methods to reduce or control the removal of stream banks by water action. See JETTY, RETARD, REVETMENT, RIPRAP.

stream bed erosion The detachment and removal of material from the bed of a stream by the movement of water, pebbles, and boulders. See LOAD, SOD STRIPS.

225

stream flow depletion The annual amount of water flowing into a basin or valley, minus the amount that flows out of basin or valley; usually less than CONSUMPTIVE USE of water, which in addition includes rainfall and water from underground sources.

stream flow routing Process of determining hydrograph at a point downstream from an upstream hydrograph aided by local knowledge of inflow and storage considerations. See FLOOD ROUTING.

stream gauging Measuring the area of cross section in a stream channel and the velocity of water flow to determine discharge; usually taken over a period to determine the trend or changing amounts of discharge. See GAUGING STATION.

stream gradient (or **slope of stream**) The general slope of the water surface or bed of a stream. See SURFACE PROFILE.

streamline A line following the direction of flow of a stream at all points along its course at a given time. See STREAM PROFILE.

streamline flow (or **laminar flow**) A flow of water which is steady and continuous and the streamlines more or less parallel at all points. See CRITICAL VELOCITY, TURBULENT FLOW.

stream loads See LOAD.

stream profile (or **thalweg**) A curve indicating the elevation at key points along the course of a stream from source to mouth; a line following the deepest part of the bed of a stream. The stream is assumed to be straightened out on plan. In general, a river has the steepest slope nearest its source or head, and least slope at its mouth, but in between there may be many irregularities.

streams (fault control) Streams controlled by fault lines. The fault may form a scarp which would act as a confining bank, or form a zone of weak resistance to erosion which is sought by streams for their channels. In these cases the stream flows along relatively straight and definite lines and does not meander as it does normally. Fault control of streams show up clearly in aerial photographs. See DRAINAGE PATTERN.

stream sections (1) See CROSS SECTION (OF STREAM). (2) Sections along hillside streams showing the rock exposures in the channel. Geological mapping in some regions consists largely of drawing sections along streams and connecting up identical beds.

strike (or **level course**) A line on a rock stratum at right angles to the full dip or slope of the stratum; a line drawn on a map to represent an imaginary line on the rock or ground of a definite level at all points.

stripping A water engineering contract item applied to topsoil removal which is not required for re-use, and the stripping of borrow areas and quarry areas of unsuitable surface materials overlying usable materials. The unit rate includes all operations in getting the material to the disposal area. Measurement is in CUT YARDS (0.765 m³).

structural analysis (or **tectonic analysis**) An investigation of the geological features shown on a map and making deductions as to the rock succession, the rock structures, and the forces active during the period; forms part of all hydro-geologic investigations.

structural trap A geological structure in which the porous and non-porous rocks are so arranged as to be favourable for the accumulation of water. See FAULT TRAP.

structure (1) Rock features such as lamination, bedding, current bedding, etc.; in soils, the grouping of the particles and the manner of breaking, such as crumb structure, block structure, platy structure, and columnar structure. (2) Any artificial construction or works to use, conserve or store water or to reduce scour.

structure bores Boreholes drilled to determine, broadly, the geological structure of an area. In the case of water, the position of the water-table and lines of faulting and folding with reference to aquifers and drainage areas are important.

structure closure (or **closure**) The elevation in metres, feet or other units, of the highest point of a flexure above the lowest 'closed' contour. Closure, nose, trap, fault, and related structures are important to the water-supply engineer.

structure contours Contours drawn on the upper surface of a rockhead, aquifer, or other deposit and used to depict its form and depth. The information for drawing structure contours is obtained from maps, boreholes, wells, springs, and perhaps geophysical surveys.

stuffing box A cylindrical recess compactly filled with hemp or similar packing material to prevent leakage of water from a pump. See GLAND.

stumping off (Australia) See PREVENTATIVE SNAGGING.

sub-artesian water ARTESIAN WATER which the pressure causes to rise in a well but not to overflow at the surface.

sub-critical flow A velocity of flow which is less than one the recognised critical values. See CRITICAL DEPTH, CRITICAL VELOCITY.

sub-glacial stream A stream formed by melting ice near the termination or SNOUT of a glacier. The water seeps down through crevasses or cracks to the bottom of the glacier.

sub-humid A climate or region where the moisture content is below that under humid conditions but normally still adequate for the production of many crops without irrigation. See SEMI-ARID.

sub-irrigation Irrigation in which measures are taken to control drainage discharge and to elevate the water-table locally to or nearly to the root zone. See SUBSURFACE DRAINAGE CHECK.

submarine blasting Blasting or the detonation of explosive charges under-water to break rock during dredging or in wet shafts or hydraulic excavations. For isolated blasting in shallow water the gelatinous explosives are in general use, namely, A.N. Gelignite 60 and A.N. Gelignite Dynamite 75. Where possible, underwater blasts are fired simultaneously using Cordtex or Submarine No. 8 electric detonators.

submarine core drill A drill capable of taking cores and samples of the deposits beneath the sea bed to assist in planning dredging operations, etc.

submarine river crossing A pipeline laid across a river below bed level.

submeander Small loops or windings formed within the banks of a stream, resulting from relatively low flows after flood has eased.

submerged float See SUBSURFACE FLOAT.

submerged intake A type of INTAKE WORKS often used to obtain a supply of water from a lake. A dredged channel is formed along the bed of the lake in which a pipe is laid with flexible joints. Loose material is placed round the pipe with hard material on top. The intake end has a bellmouth opening covered with wire mesh or screen. A variation consists of a tunnel excavated under the bed of the lake to the intake point. See EXPOSED INTAKE.

submerged orifice An opening or orifice which is wholly below the TAIL WATER level of the structure.

submerged weir A weir in which the TAIL WATER level is higher than the crest level. See FREE WEIR.

submergence ratio (or **drowning ratio**) The ratio of level of TAIL WATER to level of HEADWATER with reference to a structure crest when both water levels are higher than the crest. The distance downstream and upstream from the crest at which the water elevations are measured should be such that the structure does not influence the levels. See DOWN-STREAM TOTAL HEAD.

submersible pump Usually an electric centrifugal pump which can operate when wholly submerged in water. The motor is totally enclosed and fully protected and its position can be lowered in a shaft, well, or borehole as the water is withdrawn. Corrosion-resistant bronze and stainless steel materials are used with sealed motor stator windings. Control of the pump can be automatic and remote. Often used for dewatering shafts, mine workings, and other inundated works and also for pumping large volumes of water for irrigation from deep wells. Submersible pumps have capacities up to 450 m^3/h (100 000 gal/h) and heads up to 300 m (1000 ft). See SINKING PUMP, TURBINE PUMP.

subnormal pressure (water) Pressure converse of that in ARTESIAN WATER; related to CONFINING BEDS in which the water is pressing downward.

subpermafrost water Water underlying the permafrost. See SUPERPERMA-FROST WATER.

subsequent stream A stream whose course is determined by outcropping belts of weak or soft strata. See CONSEQUENT STREAMS, SUPERPOSED STREAM.

subsoil Broadly, the decomposed and incoherent material below the soil layers.

subsoil agricultural drain See AGRICULTURAL DRAIN, SUBSOIL DRAINAGE.

subsoil drainage The removal of subsoil water by construction of open inter-cepting ditches and the laying of drainpipes. Concrete or earthenware pipes, placed at a depth of about 0·9 m (3 ft) are often used. The pipes are butt-ended, 150 mm (6 in) or more in diameter, and laid with 6 mm ($\frac{1}{4}$ in) open joints. The distance between subsoil drains is a maximum in sandy soils and a minimum in clay. The drains are covered with stones or other suitable material. The term also applies to the removal of surplus water in sandy or gravelly subsoil by natural drainage to streams or lower levels. See CATCHWATER DRAIN.

subsoil flow (or **subsurface flow** or **seep-off**) Water which infiltrates upper soil layers and seeps laterally into streams or emerges as spring water; rate of flow or discharge of subsurface water. See SHALLOW PERCOLATION.

subsurface drainage check Controlling drainage discharge in an area to maintain the water-table at a level suitable for SUB-IRRIGATION. It may consist of a gate or other structure placed in a deep drainage ditch or covered drain.

subsurface erosion See PIPING.

subsurface float (or **double float**) A body which moves with the stream at a known depth, attached by a line to a SURFACE FLOAT which indicates its position and movement.

subsurface flow See SUBSOIL FLOW.

subsurface geology The geology of the rocks occurring below the surface; the study and interpretation of geologic data provided by boreholes and wells or by geophysical methods. At dams and other works, the subsurface geology is explored in some detail by trial bores or other means and maps and sections prepared. See SURFACE GEOLOGY.

subsurface ice (or **interstitial ice**) Ice occurring below the surface.

subsurface irrigation The use of below surface porous tiles or similar materials for irrigation.

subsurface runoff (or **storm flow, subsurface** or **interflow**) That portion of precipitation which forms a subsurface flow and is discharged into a lake or stream. See SHALLOW PERCOLATION.

subsurface storm flow See SUBSURFACE RUNOFF.

subsurface water All water occurring below the earth's surface, in the liquid, solid, or gaseous state. See GROUNDWATER.

subterranean stream A stream flowing along a subterranean passage, crevice, fissure, or large interstice. See DISAPPEARING STREAMS.

suction-cutter dredger A SUCTION DREDGER fitted with a CLAY CUTTER at the lower end of its suction pipe. The cutter breaks stiff clayey beds and gravel which are then removed by the dredger.

suction dredger A DREDGER in which the underwater mud and clayey material are removed by powerful suction pumps. See ELEVATOR DREDGER, FLOATING PIPELINE, SUCTION-CUTTER DREDGER.

suction head The head or height to which a pump can lift water, on the suction side, by atmospheric pressure; measured from the free water level in the sump. Normal atmospheric pressure sustains in a vacuum a column of water about 10·4 m (34 ft) high. Because of air leakage, etc., a pump is not placed at a greater height than about 6 m (20 ft) above the lowest water level in the sump. See STATIC SUCTION HEAD, TOTAL SUCTION HEAD.

suction pipe The pipe extending from the SUMP or SNORE to the pump chamber.

suction valve A CLACK VALVE or non-return valve placed in the suction pipe of a pump.

suction well A well from which a pump extracts water. The well is located in the river and may be part of a pumping station or may be a separate unit. The use of submersible pumps may be considered for river pumping units. See INTAKE WORKS.

sudd A floating mass of decaying trees and vegetation, etc., impeding navigation on the Upper Nile.

sudden drawdown The sudden lowering of tidal water levels; a rapid drop in the water level in a reservoir which could be dangerous in the case of earth dams.

sudden injection method Measuring discharge whereby a known weight of solute is injected into the stream at a cross section. At a second cross section downstream successive samples are taken over a period including the time taken for all the solute to pass the cross section. The samples taken should be sufficient to enable the estimation of the average concentration of solute over the sampling period. The discharge is calculated by dividing the weight of solute injected by the product of the total sampling period and the average concentration expressed as weight of solute per unit volume of water. See CONSTANT RATE OF INJECTION METHOD.

Suez Canal A canal connecting the Red Sea and the Mediterranean; opened 1869. Its length is 162·5 km (101 statute miles) and minimum width 60 m (196 ft 10 in) (navigation channel); constructed by the French engineer Ferdinand de Lesseps. See PANAMA CANAL.

sullage Mud and silt deposited by water; sewage; refuse; drainage.

sulphate-bearing soil A soil with a relatively high SO_3 content and which chemically attacks any concrete work in the ground. If the groundwater contains more than 0·1% of SO_3 a high-alumina cement is recommended for all concrete work at the site. In groundwater which contains less than 0·02% of SO_3, concrete is not affected.

sulphate test A test to determine whether a soil or its water content will have any chemical effect on concrete foundations. The sulphate in the soil is precipitated as barium sulphate and measured to ascertain the conditions.

summit canal A canal which cuts across a summit and must have water pumped into it.

sump (1) The pit, lodge, or catch basin which supplies water to a pump suction or inlet. A sump also serves as a settling basin to remove debris and dirt before the water enters the pump. Where the sump receives gravity drainage, screens may be interposed between the sump and drainage channels to remove wood, straw, and trash. It may also contain baffles to still the water and assist the settlement of sediment. (2) A receptacle sunk below the normal invert level of a trench or drain to collect and remove water. (2) A small tank adjacent to a FARM RESERVOIR or near the pump of a farm pondage. See FLOATING STRAINER, SNORE.

super-critical flow See HYPER-CRITICAL FLOW.

superficial compaction Increasing the dry density of earth or soil, by mechanical methods, in layers of about 150 mm (6 in) upwards in thickness. The compaction plant may include pneumatic-tyred rollers (often used for clays containing water), smooth-wheel rollers (for gravel–sand mixtures), and sheepsfoot rollers (for clays with low water content). Compaction is an important factor in the construction of earth dams. See COMPACTION PLANT, PROCTOR TEST.

superficial deposits Mainly the unconsolidated deposits which generally overlie the rockhead and consist of clay, sand, silt, peat, loam, or gravel or mixtures of same. Their stability and bearing capacity depend largely on the water content. See SOIL, UNCONSOLIDATED MATERIAL.

superpermafrost water Water overlying the PERMAFROST.

superposed stream A stream whose course was determined by the slopes of land forms now almost obliterated by erosion; this course is now superposed on the underlying rocks irrespective of their structure. See CONSEQUENT STREAMS.

supplemental irrigation Irrigation or watering of crops during dry periods to maintain growth; used in regions where normal rainfall is not adequate throughout the year.

supply channel (irrigation) The main channel supplying water to the irrigation area.

supply ditch A ditch from which water is drawn on to the irrigation land; it may form part of the supply channel.

supply pipe That portion of the branch supply pipe lying within the consumer's premises. See COMMUNICATION PIPE.

support, removal of See REMOVAL OF SUPPORT.

suppressed weir When the crest length of a MEASURING WEIR is the same width as the channel being measured then the nappe is not contracted in width and the weir is said to be suppressed.

surcharge depth The difference between the level of the SURCHARGE STORAGE and the crest level of the spillway or by-wash.

surcharge storage The temporary flood storage built up in a reservoir when water flows over the spillway or by-wash.

surface curve A curve showing elevation of water surface at given horizontal distances along a stream.

surface detention That part of rainfall which forms a thin layer of water over the surface when OVERLAND FLOW occurs. The depth of detention increases until discharge attains equilibrium with rate of supply to surface runoff. The term does not include DEPRESSION STORAGE.

231

surface drag (or **friction drag** or **skin friction**) The resistance to motion along the contact surface between a fluid and a solid. See HYDRAULIC FRICTION, ROUGHNESS COEFFICIENT.

surface drainage The drainage of excess surface water by constructing field ditches connected to larger open ditches, etc. See CATCHWATER DRAIN, CONTOUR PLOUGHING, DITCH, DRAIN.

surface float A small wooden or metallic object floating on the water surface and used to determine surface velocity and direction of flow. See FLOAT ROD.

surface geology The geology pertaining to the rocks exposed at or near the surface, and also the superficial deposits. The surface and subsurface geology are investigated at new dam and reservoir locations, canals, etc.

surface profile (stream) A longitudinal section or profile of the free surface of a stream or the water surface along an open channel. See STREAM GRADIENT.

surface retention See INITIAL DETENTION.

surface runoff (or **immediate runoff**) Runoff that has not passed below the surface before it reaches a stream or erodes its own channel.

surface slope See GRADIENT, SLOPE, SURFACE CURVE.

surface storage See INITIAL DETENTION.

surface tension The property displayed by the open surface of a liquid as if it was covered by a tight elastic skin; small dry particles will float, largely unwetted, on the surface. See BATHOTONIC REAGENT.

surface water All SURFACE RUNOFF, except sewerage or waste water; broadly, all water which rests on the surface of the earth and may be in liquid or solid state.

surface water drain Any pipe or ditch for the drainage of surface water. See DRAIN, SUBSURFACE DRAINAGE, SURFACE DRAINAGE.

surface water supplies Include water stored in reservoirs and lakes and that obtained from rivers and springs. Towns adjacent to a large river often take their water from that source. The waters from the lower reaches of rivers are more liable to contamination than the upper reaches. River water is filtered and may need both chemical and bacteriological purification before use. Water for London's public supplies has been taken for many years from the Thames River above Teddington. See WATER SUPPLY MEMOIRS.

surge A large wave or a great rolling swell of water; surging or heaving like a wave; a sudden increase of pressure in a pipeline resulting from closing a valve at its lower end; a sudden change in discharge of water in a channel caused by inflow of water, or by opening or closing of a gate.

surge pipe A stand pipe with an open top to release surge pressure.

surge tank An open tank to avoid loss of water during pressure surges. The tank is connected to the top of a surge pipe. Often used with pressure pipes leading to water turbines.

232

suspended-frame weir A weir with frames which can be raised and hung above the water from a bridge during floods. See FRAMED DAM.

suspended load (1) The fine particles carried by a stream, or other flow of water, in suspension for a considerable period. (2) Material collected in, or measured from, a SUSPENDED LOAD SAMPLER. See LOAD (STREAMS).

suspended load sampler A device for taking a sample of water and the particles it carries in suspension without separation of water from the sediment. See BOTTLE SILT SAMPLER.

suspended water Water in the ZONE OF AERATION. See VADOSE WATER (U.S.A.).

suspended water, gravity See VADOSE WATER.

suspension Fine particles of matter suspended or held undissolved in water. See LOAD (STREAMS).

suspension rod A rod, operated by hand, sometimes used instead of a rack and pinion in shallow water.

swabbing The moving of a plunger, fitted with a valve or solid, up and down in a borehole casing; often used to develop sand screen boreholes. It is also effective in opening choked fissures in chalk and similar rocks. See BACKBLOWING.

swallow holes Roughly circular cavities in limestone country into which streams disappear. The cavities were dissolved by percolating waters. CHEMICAL WEATHERING, KARST TOPOGRAPHY.

swamp A marsh, shallow lake, bog, or fen favourable to the growth of aquatic vegetation; soil or ground containing an excess of moisture, due in many cases to the proximity of the water-table. A swamp may be formed in the floodplain of a river, in a depression in glacial drift and in an upland area between streams. See CANALISATION, HILLSIDE SWAMPS.

swamp drainage See CANALISATION.

swash bank The part above the OUTBURST BANK of a sea bank or ridge.

swell (1) See FREE WAVES. (2) The increase in volume of a rock or soil when exposed to weathering or to the entry of water. See BULKING.

swelling pressure The force exerted by a mass of shale or clay when exposed to water in a confined space. See BULKING.

swing bridge (or pivot bridge or turn bridge) A MOVABLE BRIDGE which is pivoted at its centre and allows a vessel or barge to pass by swinging open horizontally. See TRAVERSING BRIDGE.

symbols See MAP SYMBOLS.

sympathetic retrogression The lowering of levels in one river as a result of reduction in upstream levels of river joining it.

synclinal closure (or canoe fold) An elongated basin fold with a length much longer than its width; so called because its structural contours on a map are closed; of importance in water supply problems if the rocks are favourable.

synoptic meteorology See METEOROLOGY, SYNOPTIC.

synoptic weather map A map or chart indicating the weather conditions prevailing over a district at a given time.

synthetic unit hydrograph A UNIT HYDROGRAPH prepared for a drainage basin which is ungauged; it is based on the basin's known physical characteristics.

T

tail bay That part of a canal immediately downstream of the TAILGATE.

tailgate The gate at the downstream end of a LOCK. See HEADGATE.

tail race The channel that conducts the spent water from a turbine or water wheel. See RACE.

tail water The water or water level immediately downstream of a turbine, water wheel, or structure. See HEADWATERS.

Tainter gate (after Burnham Tainter) See RADIAL GATE.

tank An artificial pond or basin for gathering and storing rainfall (especially India); in farms, a small reservoir for liquid manure. See also STORM-WATER TANKS, WATER TOWER.

tank, balancing See EQUALISING RESERVOIR.

tank, earth (or **livestock reservoir**) An artificial pond or basin formed by an excavation and or an earthen dam across a drainage channel for gathering and storing water for livestock. See FARM POND.

ttape gauge See CHAIN GAUGE.

tectonic analysis See STRUCTURAL ANALYSIS.

telecontrolled power station A HYDROELECTRIC POWER STATION operated by remote control.

temporary bank protection The protection of river banks from erosion by methods with only a few years life, as opposed to more permanent works such as walls of stone or concrete or concrete blocks or concrete sheet piling. A temporary protection may suffice along some rivers, to be renewed periodically or until vegetation can be established. See ANCHORED TREES.

temporary base-level (stream) The water level of a lake or pondage behind a BAR into which a stream flows and erosion proceeds. After a period the lake will be loaded with sediment, or the bar removed by erosion, and a new base-level formed which may be the sea. See BASE-LEVEL, GENERAL BASE-LEVEL OF EROSION.

temporary grade The grade of a stream when it flows into a lake or pondage behind a bar. In time the lake is filled or the bar removed and the stream will renew erosion to a new grade. See GRADED STREAM, LOCAL BASE-LEVEL.

temporary hardness (or **carbonate hardness**) See HARDNESS OF WATER.

temporary stream An intermittent stream which ceases to flow during part of the year; a stream with high water levels during late winter, followed by a summer drought.

temporary wilting See PERMANENT WILTING POINT.

tender (1) A formal offer by the tenderer for supply of materials or execution of engineering works as detailed in drawings and specifications for a specified sum of money. (2) A small vessel employed to attend a larger one and for supplying her with provisions, etc.

tensiometer An instrument which measures the force which plant roots must overcome in order to take in water from the soil. Also the water required to restore the soil to FIELD CAPACITY may be calculated from the tensiometer readings. Tensiostats are more elaborate forms of this instrument which automatically turn the irrigation water on and off as required.

tensiostats See TENSIOMETER.

terminal moraine The ridge of hummocks formed at the end of a glacier as the ice melts. The stages in the recession of the ice are marked by the successive terminal moraines. See GROUND MORAINE.

terrace formation (river) The cutting of steps or wide banks in the alluvial deposits of a river valley by the meandering tendency of the current; the cuts tend to deepen with each swing and the channel may become centralised. See PLAIN TRACTS, TERRACES (1).

terrace height (farm) The vertical distance between the bottom of the terrace channel and the top of the terrace ridge at adjacent points.

terrace interval (farm) The vertical or horizontal distance between corresponding points on two adjacent terraces. In staking terrace lines for construction, the vertical interval, or the difference in elevation between stake lines, is generally used. For the uppermost terrace, the interval is measured from the top of the field to the centre of the terrace.

terrace outlet channel (farm) The channel which takes the discharge from one or more terraces and conveys the water from the area; usually covered by dense perennial grass to minimise erosion.

terraces (1) Remnants of gravels deposited by rivers when flowing at higher levels than at present. (2) Ridges of earth or any embankments constructed across a sloping ground to spread runoff and minimise soil erosion. See BENCH TERRACE, BROAD-BASE TERRACE, FLOODPLAIN TERRACE, GRADED TERRACE, LEVEL TERRACE, NARROW-BASE TERRACE, NICHOLS TERRACE, RIDGE TERRACE, RIVER TERRACES.

terrace system (farm) All the terraces which occupy a field or slope and are connected to one or more outlet channels.

terrace width The horizontal distance from the high water line on the upper edge of the channel to the foot of the lower slope of the terrace ridge.

terracing See CONTOUR PLOUGHING.

Terzaghi Dr Karl von (1883—1963) The founder of modern SOIL MECHANICS.

test bore See TRIAL BOREHOLES, TRIAL PIT.

test bore pattern See BOREHOLE PATTERN.

tested capacity (of well) The maximum rate of yield by pumping from a well without depleting its water supply. The capacity is determined by tests over a specified period.

test pit   Used in U.S.A. for TRIAL PIT.

tetrahedron   A four-sided solid, especially a triangular pyramid. Steel-framed skeleton tetrahedra were used in building the Génissiat dam (1945) in the Rhone Valley to break the force of the water.

tetrapod   A precast concrete block consisting of 4 symmetrically spaced truncated cones. Developed by French engineers, tetrapods were used successfully in the construction of breakwaters in Morrocco, e.g. 360 reinforced concrete tetrapods weighing 15 t were used in the Casablanca breakwater in 1951 and 1250 tetrapods each weighing 25 t were placed in the Safi breakwater (Figure T.1).

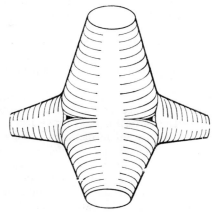

Figure T.1.   Tetrapod

thalassography   The science of the oceans and marine life. See OCEANOGRAPHY.

thalweg   See STREAM PROFILE.

thermal spring   A SPRING fed by hot groundwater and usually mineralised. In New Zealand, deep boreholes have tapped large reserves of hot water. In one the temperature was 306 °C. The water is being considered for use in a geothermal power station. See HOT SPRINGS.

thermograph   A self-registering thermometer.

thermo-osmosis   The natural movement of soil water from a relatively warm part towards a cooler part.

Thiessen polygon   A polygonal figure formed around rain-gauge stations by perpendicular bisectors of lines which connect the stations on maps.

thin core dams   Earth dams with a central impervious core width of about 15 to 20 % of the water height.

thin-plate (square-edged) weir   A thin-plate vertical weir so shaped that the sheet of water passes clear from the crest. See TRIANGULAR NOTCH.

Thornthwaite index   See MOISTURE INDEX.

three-throw pump   A pump consisting of three single-acting rams, positioned side by side, and operated by cranks set at angles of 120°. The separate suction and delivery pipes join to form common suction and discharge pipes to which air vessels are attached.

threshold velocity   The minimum velocity at which wind commences to detach and move particles of soil in a given area and under specified conditions. See WIND EROSION.

thunderstorm   A local STORM, usually of short duration, with thunder and lightning and often heavy rain or hail. See HAILSTORM.

tidal bore   See TIDAL CURRENT, HYDRAULIC BORE.

tidal current   The upstream current generated in narrow channels and shallow estuaries with the rise of the tide. The tidal current often advances with a steep wave at its front termed a BORE. The bore frequently disappears in waters or channels which have been deepened by dredging, etc. Occasionally, a similar wave may advance downstream as a result of a cloudburst or sudden release of water by removal of natural barriers in the valley. In the Mersey estuary (U.K.) the tidal current advances at about 13 km/h (8 miles per hour).

tidal dock   A dock which has no gates, so that the water level inside is the same as that of the outside tides.

tidal flood current   The tidal current advancing towards the shore and along shallow estuaries and tidal rivers.

tidal flood interval   The time interval between transit of moon over meridian of a place and the following strength of flood.

tidal flood strength   The tidal flood when current is at peak velocity.

tidal lag   The delay between high tide, or low tide, in an estuary and the corresponding highest or lowest level of the water-table in the area.

tidal marsh   A marsh located within the influence of tides and generally flooded during high tides; often traversed by interlacing channels and containing salt-tolerant vegetation such as rushes and grasses.

tidal range   Difference in altitude between high and low tidal surface.

tidal range, great diurnal   See GREAT DIURNAL RANGE OF TIDE.

tidal range, mean   See MEAN TIDAL RANGE.

tidal river   A river within the influence of tides; periodically rises and falls or flows and ebbs. See TIDAL CURRENT.

tidal surge   A rapid rise of sea level and water movement, rather like the bore of a tidal current. When superimposed on spring tides the flooding and damage is often extensive. See EQUINOCTIAL TIDES.

tidal water level   The height or elevation of a tidal surface at a specified stage of the tide. See HIGHER HIGH WATER, HIGHER LOW WATER, HIGH WATER, LOWER HIGH WATER, LOW WATER.

tidal wave   (1) In popular usage, any abnormal ocean wave of some magnitude; for example, one attributed to an earthquake. (2) In astronomic usage, a periodic variation of sea level as a result of gravitational attraction of sun and moon. (3) A STORM WAVE.

tide   The periodic rise and fall of the sea due to attraction of moon and sun; the swell of water thus generated moves round the earth as the latter rotates. See SPRING TIDES.

tide, falling   See EBB TIDE.

tide gate   A swinging gate which excludes water at high tide and allows drainage at low tide; placed below high tide level on the outside of a drainage conduit from a field.

tide gauge (or **tide predictor**)   A device with which it is possible to predict the tides in any known set of channels.

tide, rising   See FLOOD TIDE.

tie bar (or **tombolo**)   A deposit or bank of sand connecting the mainland to an island mass. On some coasts, rock reefs off the mainland have aided the formation of tie bars. See SPITS.

tile drain   Generally, concrete or earthenware pipes laid with open joints to receive and remove excess water from the soil. The internal diameter may be 100, 125, or 150 mm (4, 5, or 6 in) with a wall thickness of about a twelfth the inside diameter. See SUBSOIL DRAINAGE.

tilt dozer   A BULLDOZER to which a large degree of tilt can be applied to either side. See GRADER, SPREADER (2).

tilting gate   A spillway crest gate hinged at the bottom or top and automatically rising and falling with variations in water level; the gate is counterbalanced by weights.

timbering   The operation of setting timber supports in pipe trenches, deep excavations for foundations, large water wells, and bank protection works. The timber pieces used include poling boards, walings, struts, runners, wedges, etc.

timber sheet piling   A quick method for checking scour on a river bank. The timber piles are usually driven in a line a little outside the eroding bank and then timber sheeting fixed horizontally from pile to pile or driven vertically against horizontal walings attached to the piles. Shingle or stone is packed in the space between the sheeting and the bank. The bank may be planted with strong-rooting grasses. Timber sheet piling presents some problems such as underscour and the danger of attack by MARINE BORERS. See CONCRETE PILING.

time lag   Applied to (1) runoff of rainfall, i.e. time between beginning of rainfall to peak of runoff; (2) discharge or water level, i.e. time elapsing between occurrence of corresponding changes in discharge, or water level at two stations in a river; (3) snow melting, i.e. period between start of snow melt and beginning of resulting runoff. See LAGTIME.

time of concentration   The time taken by a droplet of rain falling on the most remote part of a drainage basin to flow to a given point under investigation. See BRANSBY-WILLIAMS FORMULA.

time of travel   The time a flood wave takes to travel a known distance downstream; the time taken for the passage of a float, solution, or other object between the two end points of a known distance along a stream or other open channel.

tipper (or **tipping lorry**) A vehicle, commonly a truck or lorry, with a body which can be elevated at one end, or sometimes sideways, to discharge its load. See DUMP TRUCK.

Titan crane A heavy crane, with a lifting capacity of at least 50 t, for handling and setting stone or concrete blocks when building breakwaters and also for ship construction. Consists essentially of a portal frame carrying a swing jib crane.

toe The lowest upstream or downstream part of a dam; sometimes referred to as the upstream toe or the downstream toe. See HEEL.

toe drain (or **toe filter**) A GRADED FILTER drain placed on the downstream toe of an earth dam to improve stability and as a protection against PIPING (Figure T.2). See APRON.

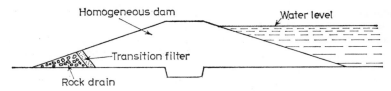

Figure T.2. Toe drain

toe filter See TOE DRAIN.

tombolo See TIE BAR.

topographical map A map showing primarily the surface features—natural and artificial—of an area. It shows seas, lakes, ponds, streams, canals, swamps, etc. Frequently used for plotting or sketching geological data and the position of wells, springs, and water-table contours. See WATER PROSPECT MAP.

topographic divide See DIVIDE.

topset beds See DELTA.

topsoil cover A layer of soil over an embankment, by-wash, or other surface to reduce erosion, minimise surface cracking, and prevent slaking of the surface material. A cover may be placed over an entire bank to a depth of 230 to 305 mm (9 to 12 in) and grassed (Figure O.1). See BANK MATERIALS, SUPERFICIAL COMPACTION.

top-soil replacement An earth dam embankment contract term: the replacement of soil on dam embankment, disposal areas, beautification areas, etc. Measured in square yards or square metres.

top water level (T.W.L.) See FULL SUPPLY LEVEL.

tornado A violent storm of limited extent but of great intensity; characterised by winds of hurricane force, rotary motion, heavy rain or hail, thunder and lightning, and often a funnel-shaped cloud; of short duration, not exceeding an hour or two, but causes much damage. Tornados occur especially in the U.S.A. from April to July and also W. Africa at the beginning and end of the rainy season. See WATERSPOUT.

tortuous flow See TURBULENT FLOW.

total capacity (of well) The maximum rate of yield by pumping after the water stored in the well has been withdrawn; the yield by pumping when the water level in the well is lowered to the intake point. See ECONOMIC YIELD.

total delivery head The sum of the static delivery head and the frictional head loss in the delivery pipe of a pump. Sometimes described as the DYNAMIC DELIVERY HEAD.

total evaporation See EVAPO-TRANSPIRATION.

total hardness of water Water hardness which includes the carbonates (TEMPORARY HARDNESS) and the sulphates (PERMANENT HARDNESS). See HARDNESS OF WATER.

total head (pump) The sum of the delivery pressure head, static pressure head, suction head, and friction head.

total head (channel) The head obtained by adding the height of the free surface above the horizontal datum and the velocity head based on the mean velocity at the given point.

total head gradient Represents the total head difference per unit horizontal length along the stream.

total head line The line or curve obtained by plotting the total head against distances measured along the stream.

totaliser A type of RAIN-GAUGE used in isolated or hilly areas where daily measurements would be difficult or inconvenient.

total moisture The moisture content of soil or other material as sampled.

total porosity Sometimes used to include both CAPILLARY POROSITY (the small voids which hold water by capillarity) and NON-CAPILLARY POROSITY (the larger voids which will not hold water by capillarity). A soil or rock with high total porosity is porous or 'open', while a soil with low total porosity is non-porous or 'dense'.

total runoff The sum of the SURFACE RUNOFF and SHALLOW PERCOLATION. See DEEP PERCOLATION.

total suction head The sum of the static suction head and the frictional head loss in the suction pipe of a pump. Sometimes described as the DYNAMIC SUCTION HEAD.

towed grader See GRADER.

toxic water Water containing poisons and tending to give a crop yield which is below normal. See ACID WATER, pH VALUE.

trace of rain See PRECIPITATION, TRACE OF.

tracer (groundwater) A dye, salt, or other substance to trace the movement of water. The tracer is introduced into the groundwater at one point and the time elapsed before it appears at a location down the groundwater gradient is noted. An ideal tracer is not absorbed by the porous material, does not react with the natural water, is not present in the natural water, can be detected in minute concentrations, and is available cheaply.

tract (stream)   One of the three major stretches or sections of a river from source to mouth. The MOUNTAIN TRACT is along its source in the mountains, followed by the VALLEY TRACT, and finally the PLAIN TRACT where the river approaches its mouth (Figure M.3).

traction (streams)   See LOAD (streams).

tractive force (water)   The force exerted by flowing water on the bed of a stream and in a direction tangential to it.

tractor-driven pump   A mobile pumping unit in which the pump is driven through the power take-off of a tractor. Often used in farm irrigation schemes where the water supply enables the pump to be mobile. A trailer is used for carrying the equipment.

training wall   A wall built along or near the sides of a river or estuary to confine the waters within the channel. A training wall consisting of a shingle embankment may be satisfactory, provided the current does not undermine it. A more permanent structure is made from piles and sheeting or a bank of gravel and earth protected on its river side by anchored trees and willows. Another type consists of stonemesh gabions placed between two rows of old rails used as piles. There are many modifications. See LOG TRAINING WALL, REVETMENT.

training works   All constructions for the regulation and training of rivers, improvement of flow characteristics, and control of scouring and silting. See GROYNES, LEVEE.

transmissibility   See TRANSMISSIBILITY COEFFICIENT.

transmissibility coefficient   The product of thickness of saturated portion of aquifer and the FIELD COEFFICIENT OF PERMEABILITY. Also called TRANSMISSIBILITY, TRANSMISSIVITY.

transmission coefficient   See COEFFICIENT OF PERMEABILITY.

transmission constant   See COEFFICIENT OF PERMEABILITY.

transmissivity   See TRANSMISSIBILITY COEFFICIENT.

transpiration   The emission of watery vapour by living plants into the atmosphere. The energy which converts the water in plants into the vapour emitted by the leaves is obtained mainly from the sun. Hence, practically all transpiration occurs during the hours of daylight. See VEGETAL DISCHARGE.

transpiration, potential   See POTENTIAL TRANSPIRATION.

transpiration ratio (or **water-use ratio**)   Relation between weight of water escaping from a living plant (roots usually excluded) during its season of growth, to weight of dry material produced.

transpiration stream   Applied to the movement of water from the soil into the plant roots, passing up the stems and then out through the leaves as vapour.

transportation (sediment)   The movement of fine material and coarse particles by stream action or waves. See LITTORAL DRIFT, LOAD, LONG-SHORE DRIFT, SEDIMENT TRANSPORT CONCENTRATION.

transport competency The ability of a stream to transport material in suspension in terms of dimensions of particles. See CAPACITY OF STREAM, COMPETENCE (of a stream), LOAD.

transport concentration See SEDIMENT TRANSPORT CONCENTRATION.

transposition of storm The use of a storm record obtained in one area at another area within a region with somewhat similar meteorologic conditions.

transverse crack A type of crack which runs at roughly right angles to the axis of earth dam. A dangerous type of crack because it could create a line of concentrated flow through the dam. Transverse cracks are due to DIFFERENTIAL SETTLEMENT.

trapezoidal groyne A four-sided stonemesh groyne with the long sides tapering to a narrow top. The stones are built on a wire mesh mat which is carried up, as the work proceeds, on both sides of the groyne and wired at the joint. See STONEMESH GROYNE.

trapezoidal weir A MEASURING WEIR constructed from thin plate with three edges forming a trapezoidal shape in a plane perpendicular to the direction of flow.

trash rack A screen of mesh or parallel bars placed across the inlet to irrigation siphons, streams, dam outlet pipes and similar hydraulic structures to intercept floating debris and prevent damage (U.S.A.).

traveller See BONING ROD.

travelling screen (1) A canvas diaphragm in a frame used for the direct measurement of the mean velocity of water. The frame travels with the water and is shaped to fit the channel, which must be of relatively uniform section. (2) A movable screen for extracting debris from flowing water.

travel time See TIME OF TRAVEL.

traversing bridge A bridge capable of moving backwards to allow vessels to pass. See MOVABLE BRIDGE.

traversing slipway A slipway for hauling vessels, up to about 500 t weight, out of the water for repairs, etc. On the slipway, the vessel can be traversed or moved sideways to another berth, leaving the slipway free for another vessel. See SLIP.

tread tractor A robust type of locomotive serviceable over rough roads; maximum speed is about 10 km/h (6 mile/h). A trailer, with capacity up to 12·0 m^3 (16 yd^3) with bottom discharge or two-way side discharge is used with the tractor. It forms an important haulage machine at many major earth-handling sites. See POWER SHOVEL.

tree planting Planting shrubs, trees, and cuttings of same for controlling gully erosion, protecting river banks, sand dunes and windbreaks; plantings in areas to prevent excessive runoff and soil erosion and to stabilise valley slopes liable to landslides. For bank protection, strong-rooting and quick-growing plants are often used. In Australia and New Zealand poplars and willows grow rapidly under a wide range of conditions and are often used along banks and to protect groynes. The bank is often

242

sloped and shallow trenches, about 0·9 m (3 ft) apart, made down the bank. A willow pole is laid in each trench with its foot in wet ground and its head extending up the bank above flood level. The willows usually root where they touch the ground along their length. See ANCHORED TREES, POPLARS, TRIPOD PLANTING, WILLOWS.

tree retards A type of RETARD used in some New Zealand rivers consisting of trees with their butts attached to wire cables anchored to concrete blocks, stonemesh gabions, or held by tar drums filled with stone. The retards are constructed out from the eroding bank and direct the current into the desired course. See SINKER GROYNE.

trellis drainage (or **grapevine drainage**) A drainage pattern in which the streams course parallel to the inclined strata outcrops for long distances and then cut through the intervening ridges and continue along another course. See DRAINAGE PATTERN.

trench, cutoff See CUTOFF TRENCH.

trench digger See TRENCH EXCAVATOR.

trench drain See FRENCH DRAIN.

trencher See TRENCH EXCAVATOR.

trench excavator (or **ditcher** or **trencher** or **trenching machine**) A self-propelled excavator, usually mounted on crawler tracks, for cutting ditches or trenches for pipelines, drainage, or water supplies; excavation may be by a chain of buckets mounted around a wheel or by a bucket ladder.

trench, groundwater See GROUNDWATER TRENCH.

trenching machine See TRENCH EXCAVATOR.

trial boreholes Holes put down as a preliminary to an important well pumping scheme. They may be up to 0·6 m (24 in) diam. for water and smaller, not exceeding about 150 mm (6 in) diam. if only geological information and water levels are required. Chemical and bacteriological analyses of the water are usually made.

trial pit (or **test pit**) Usually applied to a small shaft excavated to test the superficial deposits and position of the rockhead and water-table. It may be circular (about 0·9 m (3 ft) diam.) and put down to depths of about 15·2 m (50 ft) or more in firm clayey ground and where water is not troublesome. If rectangular, the pit is made larger, particularly where the ground must be supported. See OPEN PIT.

triangular crib groyne A CRIB GROYNE, triangular in shape, and filled with stone. It does not usually exceed about 6 m (20 ft) in length. One side of the triangle is parallel with the river bank, and the angle between the log walls at the nose is about 45°. The walls consist of small logs placed horizontally one over the other. The log ends at the nose are cut or notched to give a locking effect. The entire structure is secured by wire, rods, or spikes. The bank behind the groyne is often sloped and grassed (Figure T.3). See CRIBWORK, RECTANGULAR CRIB GROYNE.

triangular notch (or **triangular weir** or **vee notch**) A 'V'-shaped measuring weir or notch with sides that form an angle with its apex downwards.

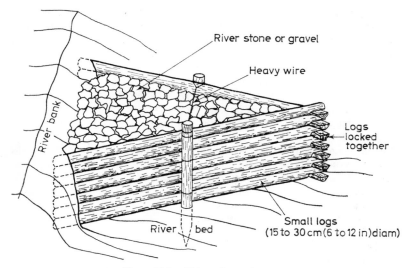

Figure T.3.   Triangular crib groyne

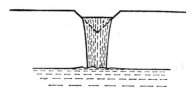

Figure T.4.   Triangular notch

It measures small flows more accurately than the rectangular-notch weir (Figure T.4).

triangular weir   See TRIANGULAR NOTCH.

tribar   A reinforced concrete block used in breakwater construction. Invented by an American, Robert Q. Palmer; consists of three cylinders so spaced that their centres form an equilateral triangle; these cylinders are joined at their midpoints by a three-pronged web (Figure T.5).

tributary   A stream that flows into a larger stream or river; it may be on the surface or underground and flow continuously or intermittently. See CONFLUENCE.

trickle irrigation   Irrigation in which special nozzles are used to cover the area from small-bore lines. A loose screw thread controls the flow of water, which drips onto the soil at a low rate and at a low pressure head. Water may be supplied direct from an overhead tank or direct from the mains. Hard water is unsuitable, as the nozzles become encrusted and

244

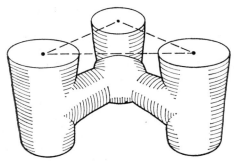

Figure T.5. Tribar

blocked. The water from each nozzle spreads downward and outward to form a cone of moist soil. See LOW LEVEL SPRINKLER IRRIGATION.

trickle tube A pipe spillway, usually between 150 and 305 mm (6 and 12 in) in diameter placed just below the bywash level of a small earth dam. It prevents erosion in the bywash by small continuous flows.

Trief process A process to provide a concrete with low heat properties for dam construction. Trief blast-furnace slag is wet-ground by the process devised in Belgium by M. V. Trief and the resulting slurry used in the concrete mixer. About 70% of the cement content can be replaced by the slag and the rise in temperature of this cement is much lower than that of Portland cement. See LOW-HEAT CEMENT.

tripod planting A recent development in New Zealand; used there to induce siltation or to check the flow of floodwater. It consists of three willow poles planted in the form of an equilateral triangle with the poles inclined inwards. They are tied together just above the ground.

tropical cyclone (or **tropical revolving storm**) Sometimes used to describe a CYCLONE in the tropics.

troughs Water vessels placed conveniently about a farm to enable stock to be adequately watered; a wooden or other channel for conveying water; any depression of a long and narrow shape for carrying water.

true velocity (groundwater) See EFFECTIVE VELOCITY (groundwater).

truncated soil profile A soil profile which has lost part or all of the A- and sometimes the B-horizon by ACCELERATED EROSION and by mechanical means; the remaining material or C-horizon consists of undeveloped parent material. The degree of truncation is indicated by comparing the eroded soil profile with virgin profile of the same area.

trunk main Usually a large-diameter pipe for conveying water supplies in bulk from one point to another within an area of supply. See AQUEDUCT.

tube well A SHALLOW WELL driven into gravels down to a depth of about 8·0 m (25 ft). It consists of a perforated pipe, usually of less than 100 mm (4 in) diam., with a pointed shoe, which is forced down into the ground.

The water is withdrawn by means of a hand pump at the surface. See COMPOUND WELL.

tubular spring An outflow of water from tubular channels in drift or soluble rocks such as limestones. The water may come from a subterranean stream and the flow may be considerable The water from tubular springs is often muddy after heavy rainfall or floods, because the water has not been filtered by percolation through porous deposits. See SWALLOW HOLES.

tubular well Almost any drilled water well of a diameter usually less than 0·6 m (2 ft) and too small for a man to descend. See OPEN WELL.

tunnel A horizontal or inclined passage for drainage or water supplies; may be provided with a continuous reinforced concrete or steel lining. A sub-aqueous tunnel may be constructed as an alternative to a bridge over a river to accommodate traffic and other facilities. See DRAINAGE TUNNEL, GALLERY, PRESSURE TUNNEL.

tunnel, infiltration See INFILTRATION TUNNEL.

tunnelling (1) The development of cracks in earthen dams and embankments under dry conditions. When rapidly brought into contact with water the cracks collapse internally to form a tunnel which could affect the safety of the construction. In dams and banks tunnelling starts at the wet face or upstream side and proceeds downstream. It thus differs from PIPING, which starts downstream and proceeds up to the wet face. Tunnelling is usually a greater danger than piping because its detection is more difficult. (2) The excavation of subways or tunnels to carry trunk mains or other works beneath rivers, through elevated ground, or where surface construction is impossible or too costly. The work is usually entrusted to a specialist contractor.

turbidity Applied to a liquid which contains suspended matter; approximate concentration of suspended matter as ascertained optically from interference to passage of light rays through a sample of the water, and compared with recorded turbidity of standard samples. See SUSPENSION.

turbine See IMPULSE TURBINE, KAPLAN TURBINE, REACTION TURBINE, TURBINE PUMP, WATER TURBINE.

turbine pump A rotary type pump with a shrouded IMPELLER with the water entering at its centre. The impeller is surrounded by a diffusion ring containing vanes which direct the impeller discharge into a circular casing which delivers into the eye of the next impeller in series. The diffusion ring converts the high-velocity discharge of the impeller into pressure head. See DIFFUSER CHAMBER, ROTARY PUMP.

turbine pump, deep well See PUMP, SINKING PUMP, TURBINE PUMP.

turbulence A state of unsteady flow in which the liquid is disturbed by eddies.

turbulent flow (or **sinuous flow** or **tortuous flow**) A fluid flow which is unsteady and the velocity at a fixed point is not constant; occurs at speeds above the CRITICAL VELOCITY of Reynolds. See STREAMLINE FLOW.

246

turbulent velocity In a particular conduit, that velocity above which turbulent flow will always exist, and below which the flow may be either turbulent or streamline, depending on flow conditions.

turfing Covering the banks of a river, irrigation channels, and river improvement slopes with grass turfs cut from other areas; used as a protection against erosion and to stabilise the slopes. See REVETMENT.

turkey's nest tank A TANK formed within a circular continuous embankment built from material excavated from outside the storage basin. The water is entirely above ground level and used for short-term storage. See RING TANK.

turn bridge See SWING BRIDGE.

twin laterals (streams) A stream which has split and continues its course along each edge of a lava flow which has partially filled a valley. If it continues along only one margin, it is called a LATERAL STREAM.

two-stage dam A dam, wall, or barrier containing two distinct drops.

typhoon A violent hurricane or whirlwind, occurring especially from July to October in China seas. See CYCLONE.

U

udometer See RAIN GAUGE.

udomograph See RAIN RECORDER.

unconfined floodplain A broad FLOODPLAIN in which the maximum meanders of the river no longer reach the sides of the valley. Usually bedrock occurs along the scouring depth during floods and the river is regarded as old. See OLD RIVER.

unconfined groundwater Groundwater not restrained in its movement by an impervious or confining bed above or below. See ARTESIAN WATER.

unconformity A geological plane or surface along which there is a break in the stratigraphic sequence and the overlying younger rocks rest upon older rocks which do not immediately precede them in the geological succession.

unconsolidated material The more or less incoherent deposits which overlie the solid rocks in most areas and consist of river alluvium, gravels, sands, clays, raised beaches, peat-bogs, etc. The material may be interbedded or intermixed. See SUPERFICIAL DEPOSITS.

uncovered open channels See OPEN CHANNEL.

undercut slopes Steeply inclined river banks on the outer edge of meanders formed by erosion along the base and caving of ground. See CAVING.

undercutting (or undermining) The erosion or wearing away of rocks or soil along the base of cliffs or banks by stream or wave action and the progressive collapse of the undermined material.

247

underflow A flow of water under a foundation, a structure, or a layer of ice; the movement of water in an UNDERFLOW CONDUIT; the rate of underflow discharge. See PIPING.

underflow conduit A deposit of porous material, under a stream, with a flow of water which is separated from the surface stream by relatively impervious clay. In some cases, the surface water is of poor quality while the underflow is clean and can be drawn upon for local supplies. It depends largely upon the thickness of the beds separating the surface flow and underflow; regular quality tests are desirable.

underground drainage (1) Drainage in limestone and other areas where much of the precipitation passes underground along channels or swallow holes, along joints and fissures, or by percolation. In the Chalk districts, the higher lands and valleys are normally dry or perhaps occupied by streams during very wet seasons only. (2) The underground gravity flow of water to sumps or pumping stations remote from the working areas. See DRAINAGE, DRY VALLEY.

underground water See GROUNDWATER, SUBSURFACE WATER, SUBTERRANEAN STREAM.

undermining See UNDERCUTTING.

underseepage See UNDERFLOW.

undertow A below surface current of water running seaward normal to the coastline and of a pulsating character. The undertow helps to scour the submerged shelf along the shore and it also carries seaward the material eroded by the water. See BACKWASH, BEACH RIDGE.

underwater apron A STONE RIPRAP which is carried as a layer or apron well out on to the bed of the river. If bed scour occurs, this extra stone will subside until it roughly beaches the new slope and thus safeguards the main slope above. This is the 'falling apron' principle, used widely in India for protection of river banks. If the river is deep, the stones may be dumped from a barge along the required lines. See STONEMESH MATTRESS.

uniform flow A steady fluid flow. See STREAMLINE FLOW.

uniform grade channel See VARIABLE-GRADE CHANNEL.

uniformity coefficient Equals size of aperture through which 60% of the sand by weight will pass, divided by size of aperture through which 10% of the sand by weight will pass. See EFFECTIVE SIZE.

unit flow, peak See PEAK UNIT FLOW.

unit graph See COMPOSITE UNIT GRAPH, DIMENSIONLESS UNIT GRAPH.

unit hydrograph A term introduced by Sherman in 1932. He defined the unit hydrograph of a river as being the hydrograph produced by an isolated storm of a given duration and of uniform intensity over the whole drainage area, which had an equivalent runoff of exactly 1 in (25·4 mm) of rain. See SYNTHETIC UNIT HYDROGRAPH.

unit of permeability See COEFFICIENT OF PERMEABILITY.

unit weight of water The weight per unit volume of water normally equal to 1 g per cm^3 or 62·4 lb per ft^3. See WATER.

unsteady flow See TURBULENT FLOW.

unusual floods See FLOOD (return period).

uplift (1) The heaving upwards of earth due to the escape of water under high pressure from a dam or other confined area. The entry of such water into sands converts them into the quick condition. The usual remedies are the sealing of leakage fissures in the ground and drainage. See FLOW NET, TUNNELLING. (2) The upward force on the base of a dam or other structure due to water pressure in the foundations.

Upper Greensand Rocks composed mainly of siliceous sandstones and sands and well developed in Dorset (U.K.). The formation yields good-quality water of varying hardness. The most productive wells are those put down in areas where the Upper Greensand is overlain by the Chalk because percolation from the latter replenishes the storage. See LIAS.

upstream total head (over a weir) A measurement at a point upstream (depending on the type of weir) of the elevation of the total head relative to the crest. See DOWNSTREAM TOTAL HEAD, WEIR HEAD.

V

vacuum tank See PNEUMATIC WATER BARREL.

vadose water (U.S.A.) Water in the ZONE OF AERATION; sometimes used to describe HELD WATER. See SUSPENDED WATER.

valley An open depression between hills or mountains; it may or may not be traversed by a stream or river. See GLACIAL VALLEY.

valley consumptive use The total amount of water evaporated from water and soil surfaces, plus that absorbed and transpired by plants, crops, and soils upon which they grow in a valley. See CONSUMPTIVE USE.

valley floodplain The flat lowlands of a valley, which are wholly or partly covered with water during floods; the floor is usually covered with waste or debris. See UNCONFINED FLOODPLAIN.

valley glaciers Glaciers which occupy valleys and which move slowly down to lower levels. The average rate of movement depends on the slope of the valley, the air temperature, and the thickness of ice. The average movement may be about 0·6 m (2 ft) per day, although rates up to 18·3 m (60 ft) per day have been recorded in Greenland. The size of a valley glacier diminishes at lower levels until it ends in a SNOUT.

valley, hanging See HANGING VALLEY.

valley-in-valley structure A structure formed when a rejuvenated stream erodes a youthful valley in the floor of the valley in which it formerly flowed. See INCISED MEANDERS.

valley spring A spring which occurs in a valley at a point where the surface falls below the WATER-TABLE or a PERCHED WATER-TABLE or any water-bearing rock. See BOURNES.

valley storage (1) Total volume of water stored in a valley between two specified points along a river, including both the channel and the floodplain. (2) Total volume of stream in flood after it has overflowed its banks. See BANKFUL STAGE.

valley tract The channel along which a stream flows in a valley or foothill country, sometimes incised in deposits laid down at an earlier period, where the gradient is low. In this region, some deposition of sediment commences; alluvial flats appear and the river meanders, sometimes from one side of the valley to the other. The valley tract may be viewed as the intermediate stage between the MOUNTAIN TRACT and the PLAIN TRACT (Figure M.3).

valley train A deposit of englacial material beyond the glacier margin in a valley. If the material is spread out over a relatively flat surface, apart from a valley, it is termed OUTWASH PLAIN or FRONTAL APRON.

valve A device, usually in a pipe, to start, stop, or regulate the flow of water. See BUTTERFLY VALVE, CLACK VALVE, DISCHARGE VALVE, STOP VALVE, SUCTION VALVE.

valve shaft A shaft connected with a reservoir and from which the outlet pipes run; also houses most of the valve controls. The upper structure of the shaft may form an exposed platform or more usually an INTAKE TOWER.

valve, sluice See SLUICE VALVE.

valve tower See INTAKE TOWER.

vapour concentration The ABSOLUTE HUMIDITY or mass of water present in a cubic metre of air. See HUMIDITY.

vapour pressure The saturated vapour pressure is the pressure exerted by the vapour above the liquid at any given temperature; it increases with increase of temperature. See SATURATED AIR.

variability of a spring The ratio, expressed quantitatively, of the fluctuation of a spring to its average discharge over a specified period.

variable-grade channel A channel along which the slope is not uniform and usually decreases as the distance from the outlet increases. In comparison, a UNIFORM GRADE CHANNEL has a slope which is constant throughout its entire length. See STREAM PROFILE.

varves Seasonal deposits from glacial streams; clayey beds containing thin alternate layers of different particle sizes.

vee notch See TRIANGULAR NOTCH.

vegetal discharge The emission of groundwater due to the physiologic functioning of living plants. See TRANSPIRATION.

vegetated channel See GRASSED WATERWAY.

vegetative practices Applied to all water and soil conservation methods based on the use of vegetation. See AFFORESTATION, GULLY CONTROL PLANTINGS, TREE PLANTING.

velocity curve, depth See DEPTH VELOCITY CURVE.

velocity gauging See CHEMICAL GAUGING, CLOUD VELOCITY GAUGING, COLOUR VELOCITY GAUGING, SALT VELOCITY METHOD.

velocity head (or **kinetic head**) The energy per unit weight of water due to its velocity V. It is also the vertical distance the fluid must fall freely in a vacuum under the force of gravity to acquire its velocity V.

$$h_v = V^2/2g$$

where h_v is velocity head, V is mean velocity of flowing water, and g is acceleration due to gravity.

velocity in pipes In general, water velocity in a pipe should not exceed about 1·5 m/s (5 ft/s). This is desirable to keep frictional losses low and also to prevent damage by water hammer effect.

velocity measurement (integration method) See INTEGRATION METHOD (VELOCITY MEASUREMENT).

velocity of approach See APPROACH VELOCITY.

velocity of groundwater See APPARENT VELOCITY (GROUNDWATER), AVERAGE VELOCITY (GROUNDWATER), EFFECTIVE VELOCITY (GROUNDWATER).

velocity of retreat The mean velocity on the downstream side of a MEASURING WEIR or structure. See TAIL WATER.

velocity rod See FLOAT ROD.

vena contracta The minimum cross-sectional area of a jet or nappe of water beyond the orifice, notch, or weir through which it emerges (Figure V.1). See COEFFICIENT OF CONTRACTION.

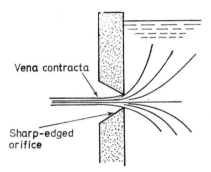

Figure V.1. Vena contracta

ventilation vents (service reservoirs) The provision of cast-iron or other ventilators in the roof, or in the walls above the water level, of a covered service reservoir. Measures are taken to prevent or minimise possible pollution of water through the ventilators.

Venturi flume A type of control flume used for measuring flow; comprises a short contraction followed by an expansion to normal width. See PARSHALL MEASURING FLUME.

Venturi meter (or **Venturi tube**) A flow meter used for closed pipes; consists of a constriction or throat followed by an expansion to normal width. The reduced pressure at the throat is measured, also the pressure upstream where the width is normal; from these points small pipes lead to gauges. The pressure difference between the points is an index of flow or discharge.

Venturi tube See VENTURI METER.

vertical alignment See REALIGNMENT.

vertical gauge See STAFF GAUGE.

vertical sand drain (or **sand drain**) A shallow hole drilled to promote the drainage of water and accelerate the consolidation of a cohesive soil. The hole is put down through a clay or silty soil and filled with sand or gravel. It should reach a permeable bed below the clay, although this is not essential. The drainage of a clay loaded by an earth dam or other heavy structure is accelerated by vertical sand drains if their spacing is less than the thickness of the clay deposit. See ARTESIAN WELL (DRAINAGE), PILLAR DRAINS.

vertical section (or **columnar sections**) A section, drawn to scale, showing the deposits penetrated in a well, borehole, shaft or major excavation. It gives details of the nature and thickness of the soils and rocks as exposed at one spot; found useful at all dam sites, cuttings, and proposed water works.

vertical velocity curve See DEPTH VELOCITY CURVE.

vibrating roller A roller for the compaction of earth or earth dams. The roller is mechanically vibrated and may be self-propelled or towed. See SUPERFICIAL COMPACTION.

victaulic joint A pipe joint which remains watertight after fixing even though the pipes are moved through several degrees.

virtual slope A graph, curve, or slope indicating the loss in pressure due to friction at any point in a system of fluid flow. See HYDRAULIC GRADE LINE.

voids (or **interstices**) The pores or small cavities between the particles in a rock or soil mass; may be occupied by water or air or both or by other liquid or gaseous substance. The term may also be applied to the spaces between the stones, gravel, or shingle in river and shore protection structures. The water velocity is reduced, resulting in the deposition of much of its silt load at and upstream of the structures. See INTERSTITIAL WATER, MICROPORES.

voids ratio The ratio of the volume of voids to the volume of solids in a sample of soil or other material. A clay contracts with increase of pressure and therefore its voids ratio may be defined at a specified pressure.

volume curve See CAPACITY CURVE.

volumetric efficiency The volume of water entering a pump cylinder for each piston stroke divided by the volume swept by the piston.

volumetric measurement A method in which the entire flow of water is discharged into a tank, of regular shape, for a measured period. Reliable means are provided for diverting or starting and stopping the flow into the tank. The method is very accurate and is used as a standard to which other measurements are referred. See DISCHARGE, GAUGING.

volute   The spiral casing of a centrifugal pump providing an area of passage which gradually reduces the speed of water leaving the IMPELLER and converting it to pressure without shock.

'V'-shaped valley   In general, a valley carved by erosion with comparatively steep sides, such as ravine, gorge, gully, or canyon. The shape of the 'V' depends upon the stage of erosion, etc.

W

wading rod   A graduated rod to which is attached a CURRENT METER for measuring the velocity in shallow streams or other channels.

wall, training   See TRAINING WALL.

warnings, flood   See FLOOD WARNINGS.

warning stage (flood)   The stage when a flood begins to cause inconvenience or damage locally near a gauge; the river may be either below or above BANKFUL STAGE. See FLOOD STAGE.

warm front   See FRONTAL SURFACE.

warp   (1) A twist or turn in a river due to an external obstruction or restraint; a MEANDER. (2) A fold in bedded rocks.

wash   The dry bed of an intermittent stream; a gully (U.S.A.).

wash boring   Putting down a drive pipe or casing through incoherent ground to bedrock with the aid of a strong water jet inside the pipe. The process may be assisted by driving.

wash load   The suspended load of a river consisting of fine particles of clay or mud derived from the erosion of the drainage area. See LOAD.

wash-outs   (1) Gutters or gullies cut by swift streams after heavy rainfall, storms, or cloudbursts; fairly common in tropical or subtropical countries. See GULLY DRAINAGE. (2) Access facilities in main pipelines for scouring, draining, or refilling.

washout valve   See SCOURING SLUICE.

waste   The part of the total output of water which is not efficiently used by the consumer; includes reservoir and pipe losses, as well as losses at the consumer's end, such as taps not turned off. See IRRIGATION EFFICIENCY, LISTENING METHODS.

wasteway   A spillway; a channel for conveying water delivered into it from a sluice, escape, or spillway.

water H_2O. Chemically pure water consists of hydrogen and oxygen combined in the ratio of about eight parts of oxygen to one part of hydrogen. See ACID WATER, FRESH, NATURAL WATER, TOXIC WATER, UNIT WEIGHT OF WATER.

water authority   A municipal body administering a system of water supply to a specified area or district. See WATER RESOURCES ACT 1963.

water-bailiff   An officer who controls and regulates water within a section of an irrigation area.

water balance The accounting for all water inputs (rainfall, snow, etc.) and all water outputs (runoff, evaporation, groundwater seepage, etc.) within a system (e.g. a catchment area). See HYDROLOGIC CYCLE.

water barrel A large barrel-shaped bucket formerly used to hoist water from the bottom of a sinking shaft or deep excavations. See PNEUMATIC WATER BARREL.

water-bearing ground The rocks in the ZONE OF SATURATION or below a PERCHED WATER-TABLE; any porous deposit containing large quantities of water seeping downwards to the zone of saturation. See AQUIFER.

water board An administrative body having authority and control of the water supplies to a district, town, or specified area.

water capacity The volume of water a soil or rock can retain; varies with the texture and composition of the rock. See MICROPORES, WATER-HOLDING CAPACITY.

water celerity See CELERITY OF WAVE.

water conservation The management, protection, physical control, and prudent use of the water resources of a district for the maximum long-term benefits to agriculture, industry, commerce, and the national economy in general. See RIVER ENGINEERING, SURFACE WATER SUPPLIES.

water consumption The amount of water consumed in a specified area. Consumption in areas with comparable conditions is often used as a guide for making estimations for new districts. The average rate of consumption for domestic purposes may be between 0·09 and 0·18 m³ (20 and 40 gal) per head per day in rural districts, and between 0·18 and 0·27 m³ (40 and 60 gal) per head per day in towns. These figures do not include water used for factories and power plants. Provision is made for a progressive increase in rate of consumption and also for probable increase in population of a district. See GROUNDWATER BUDGET.

water control The physical control of water by erection of structures for water retardation and sediment detention; improvement of channels; land practices to reduce irrigation waste and prevent depletion by seepage or evaporation. See RIVER ENGINEERING, WATER RESOURCES ACT 1963.

watercourse The path of least elevation in a valley—it is immaterial whether the path is occupied by a stream or not; any artificial or natural channel for the conveyance of water, particularly drainage. It may be a canal, stream, drainage tunnel, ditch, flume, or adit.

water creep The movement of water around the base of a structure erected on permeable deposits. See PIPING, TUNNELLING, UNDERFLOW.

water cushion See STILLING POOL.

water demand A schedule of the total quantity of water required for a specific purpose such as municipal supply, power, irrigation, or storage. See WATER REQUIREMENT.

water disposal Structures and practices for the removal of excess water from land and keeping soil erosion at a minimum. For flat ground it may include surface drains or both surface and subsoil drains. For sloping ground it may include earth dams, grassed waterways, terraces, and terrace outlet channels. See SUBSOIL DRAINAGE, SURFACE DRAINAGE, TERRACES, VERTICAL SAND DRAIN.

water divining See DIVINING, DOWSING.

water equivalent of snow The depth of water formed when snow melts. See SNOW DENSITY, SNOW MELT.

water erosion See EROSION.

water face (or water slopes) The upstream face, normally in contact with the water, of an earth dam or embankment. The face is only uncovered after prolonged dry periods or heavy draw-off (Figure A.1). See BEACHING, STONE PITCHING.

waterfalls A flow of water over a precipice; a cataract; a cascade. Caused by irregularities in the hardness of the rocks in the river or a dyke or sill in softer beds, a waterfall erodes the less resistant underlayers and undermines the more resistant ones which periodically collapse; thus the waterfall moves upstream. Falls and rapids break the navigable continuity of a river and may need blasting or even a series of locks. Again, a waterfall, if of sufficient volume, and a drop in volume, may be utilised for power generation (Figure W.1). See CATARACTS, RAPIDS.

waterfront The side of a building or structure abutting on a lake, river, or the sea; the margin of land or town on the waterside.

water gap A narrow gorge cut by a stream through a hard ridge of rock; often broad lowlands are formed in softer rocks on each side.

water hammer Produced by a rapid increase or decrease of pressure in a pipeline when the flow is stopped or checked suddenly; may be relieved in large pipes by standpipes, surge tanks, or relief valves.

water hardness See HARDNESS OF WATER.

water harvesting The art of collecting and storing water from areas which have been specially treated to increase the runoff. A very ancient practice dating back to about 2000 B.C., when it was developed in the Negev Desert of Israel. See ARTIFICIAL CATCHMENTS.

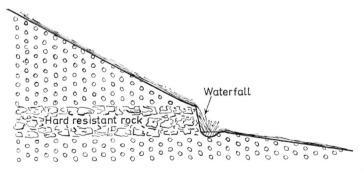

Figure W.1. Waterfall

waterhole A shallow depression or cavity in which water collects during a rainy period, or in the bed of a river otherwise dry.

water-holding capacity The minimum possible water content to which a soil sample can be reduced by gravity drainage. See CAPILLARY WATER, SPECIFIC RETENTION.

water horsepower A U.S.A. and U.K. term applied to a pump to indicate the actual rate at which useful work is imparted to the water flow and is measured in horsepower. (1 horsepower = 0·7457 kW.)

water infusion Cracking or loosening a coal seam by the injection of water under pressure by means of boreholes and pumps. Some coals are so dense that water pressure up to 28·0 N/mm^2 (4000 lbf/in^2) does not loosen them, while others are so jointed that the water passes through into the roof or floor without loosening the coal.

water level (W.L.) See MEAN WATER, STAGE, STANDING WATER LEVEL.

water lodge See LODGE, SUMP.

waterlogged Saturated with water. A depression in the ground is usually waterlogged when it is at, or below, the water-table. The condition may also result from seepage, over-irrigation, or springs.

water lowering See GROUNDWATER LOWERING.

water-meadow A meadow or grassland kept fertile by periodic flooding.

water meter An integrating FLOW METER. For measuring irrigation water there are two main types available, namely, the inferential and the rotary-piston. The rotary-piston type can be installed in any position, whereas the inferential type must be fitted in a horizontal pipeline. In general, the inferential meter requires less maintenance.

water-mill A mill whose machinery is operated by the force of water.

water of capillarity Pore water in rocks or soil above the water-table. See HELD WATER, INTERSTITIAL WATER.

water of compaction The water given off during the compaction or compression of soils or sediments with a reduction in the volume of voids. See CONNATE WATER.

water of crystallisation A fixed molecular proportion of water attached to certain minerals when they crystallise. The water can be expelled at a moderate heat as it does not enter into the inner chemical constitution of the mineral.

water of dehydration Chemically combined water which is liberated when the minerals involved undergo chemical changes.

water of infiltration The part of surface water which has entered the voids of a soil or other porous mass by INFILTRATION.

water parting See DIVIDE, WATERSHED.

water prospect map A map on tracing paper or cloth showing the position of streams, springs, depth to water-table, and probable yield of any wells put down in the area. The tracing is placed over the geological map of the same area, and on the same scale, to assist the engineer in selecting sites for wells or in solving drainage or irrigation problems.

water quality   Natural waters contain solid, liquid, and gaseous substances, in suspension or solution. Some of the points for investigation: (1) Underground waters from wells, boreholes, or springs may contain undesirable amounts of dissolved salts; those of certain metals such as zinc, copper, chromium, etc. are particularly suspect. (2) Surface waters from streams, ponds, etc. may be contaminated by sewage effluent from domestic and rural sources, and in industrial areas chemicals may also be present. (3) The amount of sodium chloride in brackish water may make it unsuitable for irrigation. Expert advice is obtained before using new water sources for irrigation, stock, or domestic purposes and its suitability is checked by chemical analysis. See BRACKISH, WATER TREATMENT.

water rate   A charge made by a water company, authority, or municipality for supply and use of water.

water regime   See MOISTURE REGIME.

water requirement (agriculture)   The total quantity of water required for agriculture; usually assessed at so much per hectare or acre or in relation to the stock population. In the U.K. the average rates commonly used are 3 to 4 gal/day per acre. Assessments based on the cattle population are usually more accurate. Typical stock requirements per head per day are: cattle 45 to 68 litres (10 to 15 gallons), horses 36 to 55 litres (8 to 12 gallons), pigs 9 to 14 litres (2 to 3 gallons), sheep 5 to 9 litres (1 to 2 gallons).

water requirement (crops)   The total quantity of water required by a plant for its growth and maturity under field conditions; includes that supplied by rainfall and any irrigation and groundwater available to the plant.

water requirement (domestic)   The total quantity of water required for domestic use, which includes hot-water systems, baths, and waterborne sewerage. In the U.K. the average figure for rural areas is between 0·14 and 0·18 m³ per head per day (30 and 40 g.h.d.). The quantity is usually estimated on the basis of population statistics and consumption is steadily increasing.

water resources   The estimated average yield of water from a given catchment area; the availability of surface water or groundwater. See GROUND-WATER BUDGET.

Water Resources Act 1963   This U.K. Act affects all municipal, civil, and mining engineers. It provides for the establishment of River Authorities and an advisory Water Resources Board. It abolished the new River Boards and transferred their powers and functions to new River Authorities, together with the functions of certain other bodies. The River Authorities have power (a) to control the abstraction and impounding of water, (b) to impose charges for licences to abstract or impound, and (c) to secure the protection and proper use of inland waters and water in underground strata. See AUSTRALIAN WATER RESOURCES COUNCIL (Appendix), WATER SUPPLY MEMOIRS.

water-resource system   A term sometimes applied to the entire system which provides water in sufficient quantities, and at reasonable rates, for irrigation, power, domestic, industrial, and other uses. A system would include the dam, reservoir together with its catchment, and all necessary works, machines, and devices. See SIMPLE STORAGE SYSTEM.

water-retaining capacity See SPECIFIC RETENTION.

water rights The legal rights possessed or acquired by an individual to the use of water. See APPROPRIATED RIGHTS, PRESCRIBED RIGHTS, RIPARIAN RIGHTS.

water-sediment mixture See SEDIMENT WATER.

watershed (or topographic divide) See DIVIDE. (In U.S.A. the term is applied to the catchment area.)

watershed leakage (U.S.A.) The flowage of subsurface water from one drainage area to an outlet in a neighbouring drainage area or directly to the sea.

water slopes See WATER FACE.

watersplash Part of road submerged by pool or stream. See IRISH BRIDGE.

waterspout A revolving tornado-like cloud descending towards the sea and sucking up a mass of water, the whole forming a whirling column connecting the cloud and sea. See TORNADO, WHIRLPOOL.

water spreading The construction of terraces across sloping land to spread and control runoff and to retard soil erosion; water control for irrigation or for subsurface storage and later extracted by pumps for irrigation. See WINTER IRRIGATION.

water stage recorder A device which gives a continuous record of the height of water at a point. The level is marked on a clock-operated chart by a float-actuated pen. See WIRE WEIGHT GAUGE.

water supply The conserving, pumping, and piping of drinkable water, usually by a public authority to consumers. See SURFACE WATER SUPPLIES, WATER RESOURCES ACT 1963.

Water Supply Memoirs Publications of the Geological Survey (U.K.). They are arranged by counties and describe the water supplies from underground sources and lists of wells are included. Address: Geological Survey and Museum, London, S.W.7. See BRITISH RAINFALL.

Water Supply Papers Publications of the United States Geological Survey. They contain information and discussions dealing with groundwater problems.

water-table The surface of the ZONE OF SATURATION, or other body of groundwater. It follows in a flatter form the profile of the land surface; it is nearer the surface in valleys and deeper along hills and elevated ground. The water-table will fluctuate as a result of natural or artificial causes. The main natural causes are rainfall, floods, barometric changes, and tides. The main artificial causes are well pumping and dams. See CONE OF DEPRESSION, MAIN WATER-TABLE, PERCHED WATER-TABLE.

water-table contour plan A plan showing contour lines on the water-table. The depths are obtained from wells, boreholes, springs, and sometimes by geophysical methods. The plan is useful for estimating the depth of proposed water wells and also when investigating water supplies and drainage schemes. Water-table contours are available for large areas of the Chalk formation of England.

water-table gradient The slope or inclination of the water-table, from the horizontal plane, at a given place and in a specified direction; may be obtained from the water-table contour plan. See PHREATIC WAVE.

water-table levels The levels at which the water-table is encountered in trial pits, boreholes, valleys, and deep excavations.

water-table, main See MAIN WATER-TABLE.

water-table, perched See PERCHED WATER-TABLE.

water-table profile A vertical section of the water-table in a specified direction; constructed from a WATER-TABLE CONTOUR PLAN or from levels in trial pits, wells, boreholes and valley springs.

watertight barrier (earth dam) See CORE (2), CUT-OFF TRENCH.

watertight seal (irrigation) A synthetic rubber seal devised to reduce leakage through flow control structures. It ensures complete watertight conditions between the vertical movable gate and the main structure. It is estimated that the seal has resulted in a saving of 123·35 million m³ (100 000 acre feet) of water in the irrigation systems of Victoria, Australia. It was developed by E. P. Robinson, an Australian agricultural scientist.

water-to-earth ratio See STORAGE/EXCAVATION RATIO.

water tower Generally a cylindrical reinforced concrete tower or tank up to a capacity of 3375 m³ (750 000 gal) or even more. The tank is elevated above ground level and serves the same purpose as a SERVICE RESERVOIR.

water treatment The purification and treatment of a water supply to ensure that it is drinkable and suitable for domestic and industrial use. It may include (1) neutralising acid water, (2) softening water of excessive hardness to make it suitable for steam boilers and washing, and (3) removal of iron or other undesirable substance. See ACID WATER, HARDNESS OF WATER, pH VALUE, TOXIC WATER, WATER QUALITY.

water turbine An arrangement in which water power rotates a wheel or prime mover which in turn is coupled to an alternator to generate an alternating current. See TURBINE.

water-use ratio See TRANSPIRATION RATIO.

waterway A navigable channel; a watercourse. See FARM WATERWAY, RIVER, STREAM.

waterway stabilisation structure See GULLY STABILISATION STRUCTURE.

water well (or **water bore**) A well put down to yield a supply of water; a well put down to yield groundwater information; a well put down to replenish groundwater. See SHALLOW WELL.

water-wheel A wheel formerly used for raising water by means of boxes or buckets fitted on its circumference; a wheel to drive machinery with water as the motive power.

waterworks planning Investigations and studies to ensure that the water resources of a region are exploited to the best advantage to avoid wasteful and haphazard schemes. For this purpose the region is divided into areas or units which will be convenient for water supply purposes. Planning may be undertaken for water requirements up to 30 years or even longer. See IMPOUNDING SCHEMES (PLANNING).

water year   A water record extending over a continuous period of twelve months. In the U.K. the year is from 1 October to 30 September.

wattle work   See HURDLE WORK.

wave   A moving ridge or swell of water between two depressions; generated by wind action in the sea, lake, or other body of water; in relatively shallow water, the ridges or rollers follow each other at regular intervals and successively break on the shore. See FORCED WAVES, FREE WAVES.

wave base   The lower limit or depth, for any particular series of waves, at which disturbance of the water ceases; erosion of the bed proceeds if the water is shallower than the wave base. See BREAKERS.

wave-built terrace   Sometimes applied to the upper coarser portions of a beach along which the waves have thrown the pebbles up into low ridges just above mean high-water level and parallel to the shore line. See OFFSHORE TERRACE.

wave-cut terrace   The shelf covered by shallow water at the base of a SEA CLIFF. In some areas, the terrace has been elevated to form a bench above sea level, as on the coast of southern California and elsewhere.

wave erosion   Erosion by waves, which includes the abrasive action of solid particles in the water and the scouring effect of the water itself. Wave erosion does not necessarily include transportation of the material. Shore currents and wave action are also active on inland seas and lakes. See SHORE EROSION.

wave height   The height from the trough to the crest; generally ranges up to about 9·0 m (30 ft) and it becomes considerably greater in heavy seas. See HAWKSLEY'S FORMULA, STEVENSON'S FORMULA.

wavelength   The distance from crest to crest; varies from 61 m (200 ft) to about 183 m (600 ft) for FORCED WAVES in the open sea; the length of FREE WAVES may be greater.

wave of the water-table   See PHREATIC WAVE.

wave pressure   The pressure exerted by waves along the shore may vary from about 24 to over 240 kN/m² (500 to over 5000 lbf/in²). The latter figure would apply to winter gales. Experiments with models have shown that waves breaking against a wall produced shock pressures ranging up to 190 kN/m² (4000 lbf/in²). See TIDAL SURGE.

wave refraction   The swinging around of an originally straight wave on approaching an indented shore line and becomes parallel with it. In the case of FORCED WAVES the effects are modified.

wave wall   See PARAPET WALL.

Wealden   The Wealden Series of the Cretaceous outcrop in the Weald of south-east England; consists of alternating beds of sand and clay. Good supplies of water are obtained in areas where the geological structure is favourable for the accumulation and replenishment of the water. The quality tends to be variable but is usually moderately soft. See UPPER GREENSAND.

weather The atmospheric state of a region at any given period, particularly temperature, humidity, cloudiness, precipitation, barometric pressure, and wind velocity. See CLIMATE.

weathering Embraces all those changes which occur in soils and rocks as a result of the action of heat and cold, wind, rain, snow, hail, exposure to air, and other atmospheric processes. See BELT OF WEATHERING, CHEMICAL WEATHERING.

Weber's number A number giving the ratio between influences of inertia and surface tension in a fluid.

weep-holes Small apertures or pipes formed during construction of retaining walls, bridge abutments, aprons, or foundations to permit drainage of water and thus reduce the pressure behind the structures.

weighed filter See LOADED FILTER.

weir A concrete, stone, or stonemesh overflow wall or structure erected across a stream. It is not used to store water but to retard flow and raise the upstream water level to permit diversions and perhaps to measure the rate of flow. Weirs are sometimes constructed in series, in which case the effect of the lower weir will extend to the next one above it. See CIPOLETTI WEIR, MEASURING WEIR, SHARPCRESTED WEIR.

weir head The depth of water measured from the bottom of the notch to the water level at a point upstream of the weir. The VELOCITY OF APPROACH is not included.

weir notch The opening in a weir for the passage of water. See MEASURING WEIR, NOTCH.

weir, overflow See OVERFLOW WEIR.

well A small pit excavated or drilled from the surface to obtain water; it exceeds 0·9 m (3 ft) in diameter if excavated by hand. The usual depth in the U.K. is between 61 and 122 m (200 and 400 ft). A well may also be used for other purposes. See DRAINAGE WELL, OBSERVATION WELL, OPEN WELL, SHALLOW WELL, TUBE-WELL.

well, cavity See BOULDER WELL.

well drain See ABSORBING WELL.

well field A region or locality containing a number of wells which supply water for public, industrial, or irrigation purposes. The wells may be pumped individually or in one or more groups or batteries. See BATTERY OF WELLS.

well, flowing See FLOWING WELL.

well hydrograph A graph showing fluctuations of water level in a well; levels marked as ordinates against time as abscissa.

well interference Used when the CONE OF DEPRESSION of a well is affected by that of an adjacent well or wells and causing a decrease in its yield of

261

water. The diminution may be expressed as a lowering of water-table in metres or feet or reduction in yield.

well, inverted   See INVERTED WELL.

well log   A description of the rocks passed through in a well; similar to a borehole log but usually not so detailed. See DRILLER'S LOG.

wellpoint   A shallow well with a pump to drain a water-bearing soil around or along an excavation. A wellpoint strengthens the ground and reduces the pumping required inside the excavation. It consists essentially of a tube of about 50 mm (2 in) diameter which is driven into the ground by jetting. The tube is fitted with a close-mesh screen at the bottom and connected through a header pipe to a suction pump at the surface. For a major job, a number of wellpoints are sunk and connected to a common header pipe. Fine sands cannot be drained by wellpoints. See GROUNDWATER LOWERING.

well pressure   The natural pressure of water in a well; it may be sufficient to cause the water to rise to the surface and form a FLOWING WELL. Well pressure is not related to the depth of the well or of the water. See ARTESIAN WELL.

well record   A brief account of the history of a well and all the available data regarding yield of water, pressure, depth, and rocks penetrated. See WELL LOG.

well screen   A screen or strainer placed in a well when pumping water from a loose, gravelly deposit. The screen excludes solid particles over a certain size but allows fine sand to enter with the water and be removed. The large particles accumulate around the well to form a natural gravel screen. The screen is made from non-ferrous metal and the two main types are tube strainers with milled slots and wire-wound strainers. See BONDED GRAVEL SCREEN, GRAVEL FILTER WELL.

well section   A vertical section showing the rock deposits passed through in a well. The scale used may be 1:100 (10 ft to one inch) or more, depending on the depth and the details to be shown. See WELL LOG.

well sinker   A man skilled in excavating shallow wells, generally with pick and shovel, to yield a local supply of water. Usually a local man, he is familiar with the superficial deposits of the district and also the general lie of the water-table.

well water   The water as pumped from a well or obtained from a FLOWING WELL.

wet-bulb thermometer   An ordinary thermometer with its bulb covered with muslin which is kept permanently moist with pure water; used in conjunction with dry bulb as HYGROMETER.

wet dock   A dock within which the water is maintained at high-tide level by the dock gates, which are opened only at high tide. See DRY DOCK.

wet gap   A military term used for obstructions involving water, such as streams, ponds, rivers, etc.

wet line correction   See AIRLINE CORRECTION.

wetted perimeter   The length of surface in a stream, conduit, or channel in contact with the water, measured at right angles to the flow. See HYDRAULIC RADIUS.

wet well   See DRY WELL (2).

wharf   A BERTH or platform of timber, stone, or concrete constructed parallel to the waterfront, where ships may be moored for loading or unloading. A SOLID WHARF acts similarly to a retaining wall and holds back all the earth behind it, while a OPEN WHARF does not retain the earth behind it, and is supported on piles of steel, precast concrete, or sometimes timber forced down into the bed.

whirling hygrometer   See HYGROMETER.

whirlpool   A large circular eddy in a stream or the sea, often of great power; may be caused by an obstruction or spur projecting into the water or channel. See MAELSTROM, WATERSPOUT.

whistling buoy   A BUOY in which wave action activates an attached whistle to give audible warning.

whole-tide cofferdam   See FULL-TIDE COFFERDAM.

wicket   A small door or gate, especially one forming part of a larger one in any hydraulic structure; a sluice in a lock gate.

wicket dam   A movable dam consisting of shutters or wickets which can be revolved about a central axis.

width (or breadth (stream))   The linear dimension measured at right angles to the direction of flow. See CROSS SECTION.

willowbox groynes   Usually small groynes, similar in construction to FASCINE GROYNES, and used for local bank protection. They consist of willow fascines or willow spars and the interior is filled with shingle. The top is commonly covered over with a layer of fascines.

willow groyne   A type of PERMEABLE GROYNE used successfully in New Zealand where the stream bed is favourable for the growth of willows. The groyne is made by wiring green willow stakes to horizontal cables attached to concrete blocks buried in the river shingle. In time, the stakes develop and form a thick barrier of willows. There are structural variations, depending on local conditions and objectives.

willows   Trees furnishing pliable shoots and growing in temperate climates near water. They are divided into tree willows, pussy willows, and osiers. Tree willows are used for stream stabilisation and protection; the shallow roots must be regularly cut back for river protection. Pussy willows are used on drier ground, while osiers may be used at sites where their bitter flavour gives them immunity from grazing stock. See TREE PLANTING.

wilting coefficient   The moisture, in percentage of dry weight, remaining in the soil when plants reach a stage of permanent wilting. Given by the expression

$$100 \left(\frac{d}{W} \right)$$

where d is the weight of water in the soil when permanent wilting begins and W is the weight of the soil when dry. See PERMANENT WILTING POINT.

wilting point   See PERMANENT WILTING POINT.

wind erosion   The detachment and removal of soil particles by wind action; a serious problem in some regions during dry periods, particularly where the surface layer is loose and the vegetation is inadequate to bind the soil substance. See THRESHOLD VELOCITY.

wind-wave flume   A special flume used in small-scale models concerned with waves due to wind action. See MODEL ANALYSIS.

wing dam   See SPUR.

wing trenches   Long trenches or grout curtains to make the ends of a dam watertight. They are constructed beyond the ends of the dam and continue up the valley sides for distances which depend on the geological conditions at the site.

wing wall (or **abutment wall**)   A side wall at the abutment of a bridge extending beyond the bridge to hold back earth; a side wall extending from the headwall of a dam structure downstream the length of the apron, so as to confine and direct overfall and to reduce sloughing of channel banks.

winter irrigation   The storage of water in the soil and subsoil by irrigation between growing seasons for subsequent use by plants and crops. See WATER SPREADING.

wire dam   A barrier in which the headwall is constructed from mesh wire.

wire mesh   Any wire netting or fabric used to wrap and secure stones or other material in a river or other water control structure. Small-gauge wire with small openings is usually used for light work such as securing brushwork on banks, while heavy wire with large openings is used for heavy structures such as large groynes. See STONEMESH CONSTRUCTION, STONEMESH GROYNES, STONEMESH MATTRESS.

wire weight gauge   A RIVER GAUGE consisting of a weight suspended on a wire for measuring the water level under a bridge, often with a datum marked on the bridge structure. See CHAIN GAUGE, ELECTRICAL TAPE GAUGE.

Woolf system (river training)   A system often used in Europe for the same purpose as the LOG TRAINING WALL. It employs piles and walings which induce the deposition of shingle and so build up a bank. Used sometimes to reclaim an area of land, in which case gaps are left to allow the shingle to be carried through to the area beyond. See SLOPING OF BANKS.

working chamber   The space in which men work on excavation at the foot of a pneumatic CAISSON.

working head (pump)   The sum of the TOTAL SUCTION HEAD and the TOTAL DELIVERY HEAD.

working map   A BASE MAP used during GEOLOGICAL MAPPING.

working shaft   A small pit to gain access to and to excavate a sewer, drain, or other works. After construction, the shaft is usually filled with earth or stone.

Wright, Benjamin A famous U.S.A. water and civil engineer of the 19th century. He has been designated by the AMERICAN SOCIETY OF CIVIL ENGINEERS (Appendix) as the 'Father of American Civil Engineering'. Well known for his work on the Erie Canal.

Y

yard A British and U.S. unit of long measurement. 1 yard = 0·914 m. See also METHOD OF MEASUREMENT.

yard gulley A small inlet, covered with a grating, to a drain to receive rainwater and waste water. See GULLEY.

yard trap See GULLEY TRAP.

yield Generally, the ECONOMIC YIELD of a well. The permeability of the rocks, and the probable yield of a new well penetrating such rocks, may be obtained from a short pumping test giving the different values of drawdown for successive increases in rate of pumping. The information enables the probable yield of the well to be estimated. This is a quicker method than carrying out a full-scale pumping test over a period. Sometimes applied to CATCHMENT YIELD. See OVERPUMPING.

yield of drainage basin The average annual runoff for the basin. See DRAINAGE COEFFICIENT.

young stream See YOUTHFUL STREAM.

youthful stream A stream which flows swiftly in a narrow steep-sided valley; it cuts vertically rather than laterally; its course contains many irregularities such as rapids, pools, and waterfalls. A youthful stream has a long period of degradation and silting before it attains a PROFILE OF EQUILIBRIUM. See EARLY YOUTH.

Z

zeolite process A process to soften hard water (see HARDNESS OF WATER) by the base-exchange method in which ZEOLITE is used to remove the salts. Back-flushing with brine regenerates the zeolite layer. The process is not suitable for treating water containing any chloride or sulphate of magnesium.

zeolites (1) Artificial substances used in the base-exchange method of softening hard water. (2) Natural hydrated aluminous silicates of calcium, sodium, or potassium; occur in cracks and cavities in basalts, lavas, as white or glassy crystals.

zero moisture index See MOISTURE INDEX, ZERO.

zone A layer, belt, or mass of ground or other material bounded by marginal horizons or surfaces which may be plane or curved and flat or inclined and within which certain physical, chemical, or other conditions prevail. See below.

zone, intertropic convergence See INTERTROPIC FRONT.

zone of aeration The ground above the MAIN WATER-TABLE and extending to the surface. In ascending order, the zone may be divided into the CAPILLARY FRINGE, an intermediate belt of ground and the BELT OF SOIL WATER. The zone of aeration may vary in thickness according to local conditions.

zone of capillarity The ground above the water-table containing WATER OF CAPILLARITY.

zone of saturation The mass of water-bearing ground below the main water-table. The zone may consist of incoherent material and solid rocks.

zone of weathering (or belt of weathering) The layer of superficial deposits which is subject to weathering and coincides broadly with the BELT OF SOIL WATER.

zoogloeal layer A gelatinous type layer comprising algae, suspended matter, diatoms, and micro-organisms which forms on the surface of the sand of slow sand filters. It is the filtering agent which retains the impurities in the water. Also called SCHLAMMDECKE.

APPENDIX: Organisations

Advisory Commission of Inter-Government Relations A Commission formed to arrange co-operation between local, State and Federal governments on problems of sewage and water. It may make legislative recommendations on these matters. Offices: 1701 Pennsylvania Avenue, NW, Washington DC, 20575.

American Public Power Association An Association concerned with all aspects of management and operation of public electric power systems and has a secondary interest in water research associated with these projects. Offices: Suite 830, 919 18th Street, NW, Washington DC, 20006.

American Society of Civil Engineers A body of professional engineers formed in 1852 to promote interest in all branches of civil engineering. It has fifteen divisions, five of which involve a form of water engineering, namely hydraulics, irrigation and drainage, pipelines, sanitary engineering, and waterways and harbours. Offices: United Engineering Centre, 345 East 47th Street, New York, N.Y. 10017.

American Society of Limnology and Oceanography Formed to encourage interest in all forms of marine and fresh waters. Main functions include water resources policy, water quality, staffing of aquatic sciences, equipment for limnological and oceanographic research. Offices: C/o Department of Oceanography, Oregon State University, Corvallis, Oregon 97331.

American Water Resources Association An Association to stimulate interest in the general aspects of water resources research, their planning, development and management. Undertakes water resources education and dispenses water resources information. Offices: 103 North Race Street, P.O. Box 434, Urbana, Illinois, 61801.

American Water Works Association (A.W.W.A.) An organisation concerned with the operation, management and maintenance of public water supplies systems, including water treatment techniques, water quality, distribution of water supplies, water storages and water rates. Offices: 2 Park Avenue, New York, N.Y. 10016.

Australian Water Resources Council Formed to provide a comprehensive assessment, on a continuing basis, of Australia's water resources (surface and underground) and the extension of measurement and research so that future planning can be carried out on a scientific basis. The assessment would indicate (a) areas which offer the greatest potential for absorption of population growth from the point of view of water resources, and (b) areas lacking in adequate water resources, where special measures may be needed to provide opportunities for development. Headquarters and Secretariat: c/o Department of National Development, Canberra, Australia. See WATER RESOURCES ACT 1963.

The British Non-ferrous Metals Research Association Established 1920; the national research organisation of producers, manufacturers, and users of non-ferrous metals. Many water engineers have received valuable assistance from the Association in the investigation of corrosion problems. Offices and Laboratory: Euston Street, London N.W.1. See INSTITUTION OF WATER ENGINEERS.

British Standards Institution The U.K. organisation recognised by industry and by Government for the preparation of national standards for use in connection with the products of all branches of industry and commerce, except for a few specialised substances. The Institution does not within itself initiate standardisation: it is the instrument through which standards are evolved as a result of free discussion between all the parties interested. Offices: British Standards House, 2 Park Street, London W.1. (See THE BRITISH NON-FERROUS METALS RESEARCH ASSOCIATION.)

British Waterworks Association Founded 1912; includes in its membership some 600 water authorities supplying, in aggregate, over 45 million persons in Great Britain, Northern Ireland, and certain dominions. Its primary function is the safeguarding of the interests of the water undertakers and, through them, the interests of the water consumers they serve. Offices: 34 Park Street, London, W.1. (See THE WATER COMPANIES ASSOCIATION.)

Federal Power Commission (F.P.C.) An authority empowered to license hydroelectric schemes on U.S. Government lands or waters; jurisdiction over the inter-state sale of electric and natural gas supplies; responsible for water resources planning and developing for projects associated with conventional and pumped storage hydroelectric systems. Offices: 414 G Street, NW, Washington DC 20426.

Freshwater Biological Association Incorporated 1932; its membership includes private individuals (both scientific and anglers), universities, scientific societies, water undertakers, fishing clubs, and river boards. Fundamental research is carried out on many aspects of the biology of fresh waters and the results are published in scientific journals and summarised in Annual Reports. Headquarters: The Ferry House, Ambleside, Westmorland.

Geological Society of London Founded 1807; granted a Royal Charter 1825. The senior geological society of the world, its principal objective is the furtherance of geological science in all its aspects. Headquarters: Burlington House, London, W.1.

Glaciological Society Formed to stimulate interest in and encourage research into scientific and technical problems of ice and snow in all countries. Address: c/o Scott Polar Research Institute, Lensfield Road, Cambridge, England.

Great Lakes Commission A Commission formed to safeguard the use of the water resources of the Great Lakes Basin. Its main functions include control of water pollution, protection of fisheries, prevention of shore erosion, control of navigation, regulation of water diversions from and into the Great Lakes. Offices: 5104 1st Building, North Campus, Ann Arbor, Michigan, 48105.

Institution of Civil Engineers Founded 1818 for the general advancement of non-military engineering; today the senior professional civil engineering society in the Commonwealth. Granted a Royal Charter 1828. Headquarters: Great George Street, Westminster, London S.W.1. (See CIVIL ENGINEERING.)

Institution of Water Engineers Established 1896, incorporated 1911. Membership includes water engineers to local authorities and companies, consulting water engineers and land drainage companies. Of recent years, 10 Research Groups have been established, each investigating a technical problem of practical importance to members. Offices: 11 Pall Mall, London S.W.1. See BRITISH WATERWORKS ASSOCIATION.

International Commission on Irrigation and Drainage (I.C.I.D.) An organisation to stimulate and promote the development and application of the science of irrigation, drainage, flood control, and river training in the engineering, social, and economic aspects. Headquarters: 184 Golf Links Area, New Delhi II, India.

International Commission on Large Dams (I.C.O.L.D.) Formed to encourage improvements in the design, construction, operation, and maintenance of large dams by bringing together information for study. Headquarters: 51 rue Saint-Georges, Paris 9e, France. See LARGE DAM.

International Joint Commission—United States and Canada Formed to foster co-operation beetwen U.S.A. and Canada on the use, diversion, obstruction, and pollution of boundary waters between the two countries. Offices: 1711 New York Avenue NW, Washington DC 20440.

International Boundary and Water Commision–United States and Mexico An administration established to control the diversion, distribution, and regulation of the waters along the boundary of the United States and Mexico as laid down by formal agreements. U.S. Offices: P.O. Box 1859, El Paso, Texas, 79950.

International Water Supply Association Founded 1947 at a meeting convened by the BRITISH WATERWORKS ASSOCIATION. Its objects are to establish an international body concerned with the public supply of water through pipes for domestic, agricultural and industrial purposes; to secure concerted action in improving the knowledge of public water supplies; to secure a maximum exchange of information on research, methods of supply of water, statistics, and other matters of common interest. Offices: 34 Park Street, London W.1.

National Association of Corrosion Engineers An Association formed to study all corrosion problems associated with waste water treatment and the re-circulation of cooling water, desalination, and the effects of exposure to water of metallic and non-metallic surfaces. Offices: 980 M and M Building, Houston, Texas 77002.

National Joint Industrial Council for the Waterworks Undertakings Industry

The National Council, which was set up in 1919, exists to enable organisations representative of employers and employees in the industry to meet to discuss problems of mutual interest, to endeavour to settle disputes without strikes or lock-outs, and to negotiate collectively the rates of pay and conditions of service for manual employees throughout the industry. Offices: 34 Park Street, London W.1.

National Water Well Association   Formed to improve knowledge on water well drilling, hydrogeology, and various aspects of groundwater engineering. Offices: 1201 Waukegan Road, Glenview, Illinois, 60025.

Society for Underwater Technology   Founded 1967 to promote the knowledge and practice of underwater technology. It has technical publications and brings together representatives of industries, Government departments, and academic research and organises meetings. Headquarters: 1 Birdcage Walk, London S.W.1.

Standing Committee on Water Regulations   A Committee of the BRITISH WATERWORKS ASSOCIATION and was instituted in 1919. Its work is concerned with the formulation of water regulations and bye-laws, the standardisation of water fittings and examining and reporting on new kinds of fittings and appliances. Jointly with BRITISH STANDARDS INSTITUTION, it operates a scheme whereby manufacturers are licensed to place the registered standardisation marks of the Institution and of the Committee on water fittings made in accordance with certain British standards. Offices: 34 Park Street, London W.1.

Tennessee Valley Authority (T.V.A.)   An organisation founded in 1933 to provide navigable channels and flood control on Tennessee River and its major tributaries. It also administers the construction of hydroelectric power on the rivers and is empowered to dispose of surplus electric power. Offices: New Sprankle Building, Knoxville, Tennessee, 37902.

U.S. Committee on Irrigation, Drainage and Flood Control   A Committee formed to encourage the study of all phases of irrigation, reclamation, flood control, river improvement and the economics of such schemes. Office: P.O. Box 15326, Denver, Colorado 80215.

U.S. Committee on Large Dams   A Committee concerned with the design, construction, operation and maintenance of large dams in U.S.A. Office: C/o Engineers Joint Council, 345 East 47th Street, New York, N.Y. 10017.

U.S. Water Resources Council   Formed to coordinate Federal water resources planning and action programmes; guidance of Federal policy on water resources planning; development of principles, technical standards and procedures for planning and evaluation of Federal Agencies projects. At the request of the States, the Council may establish Federal-State river basin planning commissions. Offices: 1025 Vermont Avenue NW, Washington DC 2005.

Victorian Irrigation Research and Promotion Organisation (V.I.R.P.O.)   An organisation formed in Victoria, Australia, in 1966 to bring about a more efficient use of available water supplies, to increase the productivity of irrigation farms, to emphasise the importance of irrigation in both the State and national economics, and to ensure that maximum efforts were made in the development of water resources and combat attacks on irrigation.

Water Companies Association Founded 1885 under the name of the Provincial Water Companies' Association and has a membership of 84 water companies. Its main functions are to safeguard the interests of the members; to encourage and facilitate the efficiency of water companies; to watch over, protect, and promote their interests and the interests of the public in regard to the supply of water; to encourage the development and improvement of the water industry; to study legislation affecting the water industry, etc. Offices: 15 Great College Street, London S.W.1.

Water Pollution Control Federation An organisation concerned with all aspects of water pollution control and waste-water disposal. Offices: 3900 Wisconsin Avenue NW, Washington DC 20016.

Water Pollution Research Board A Board, which operates under the aegis of the Department of Scientific and Industrial Research, is a body of scientific and technical experts 'to advise on the conduct of research on the prevention of pollution of waters, on the treatment of waters to improve their quality and on related topics'. Results are published in Annual Reports. Offices and laboratory: Water Pollution Research Laboratory, 103 Langley Road, Watford, Herts.

Water Research Association Incorporated in 1953, under the Companies Act 1948. Its purpose is to carry out fundamental research into problems of the water industry—in particular of public water supply, to undertake applied research and development relating to technological aspects of the water industry, and to provide an information and advisory service to its members. Offices: Ferry Lane, Medmenham, Marlow, Bucks.

Water Research Foundation of Australia An institution to initiate, promote, and further scientific and technological research into the development, control, and use of the water resources of Australia; to foster the work of and co-operation in research carried out by universities and other organisations in the development, supply and use of water generally.

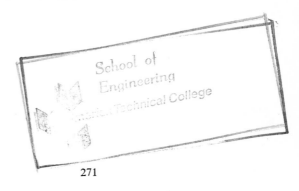